War and Tropical Forests: Conservation in Areas of Armed Conflict

Steven V. Price
Editor

War and Tropical Forests: Conservation in Areas of Armed Conflict has been co-published simultaneously as *Journal of Sustainable Forestry*, Volume 16, Numbers 3/4 2003.

CRC Press
Taylor & Francis Group
Boca Raton London New York

CRC Press is an imprint of the
Taylor & Francis Group, an informa business

CRC Press
6000 Broken Sound Parkway, NW
Suite 300, Boca Raton, FL 33487
270 Madison Avenue
New York, NY 10016
2 Park Square, Milton Park
Abingdon, Oxon OX14 4RN, UK

Published by

Food Products Press®, 10 Alice Street, Binghamton, NY 13904-1580 USA

Food Products Press® is an imprint of The Haworth Press, Inc., 10 Alice Street, Binghamton, NY 13904-1580.

War and Tropical Forests: Conservation in Areas of Armed Conflict has been co-published simultaneously as *Journal of Sustainable Forestry*, Volume 16, Numbers 3/4 2003.

Cover design by Jennifer M. Gaska

Cover photo by Steven V. Price. Munchique National Natural Park, Colombia, 1995. Sign reads, "danger mined zone."

Library of Congress Cataloging-in-Publication Data

War and tropical forests : conservation in areas of armed conflict / Steven V. Price, editor.
p. cm.
"Co-published simultaneously as Journal of Sustainable Forestry, Volume 16, Numbers 3/4 2003."
Includes bibliographical references (p.).
ISBN 1-56022-098-8 (hard : alk. paper)–ISBN 1-56022-099-6 (pbk : alk. paper)
1. Forest conservation–Tropics. 2. Forests and forestry–Tropics. 3. War–Environmental aspects–Tropics. I. Price, Steven V. II. Journal of Sustainable Forestry.
SD414.T76 W27 2002
333.75'16'0913–dc21

2002152072

War and Tropical Forests: Conservation in Areas of Armed Conflict

CONTENTS

ABOUT THE EDITOR

Steven V. Price, MF, MA, is a forestry consultant. He was the principal organizer of the "War and Tropical Forests: New Perspectives on Conservation in Areas of Armed Conflict" conference held in 2000 at the Yale School of Forestry and Environmental Studies. Mr. Price holds master's degrees in forestry and international relations and was a Fulbright fellow in Colombia during 1994-1995.

Preface

From the lowland rainforests of the Colombian Amazon to the rugged habitat of Rwanda's mountain gorillas, wars are having severe impacts on tropical forests and the communities that they sustain. Armed conflicts, and the political, economic, and humanitarian crises they provoke, often lead to unsustainable exploitation of tropical forest resources and widespread habitat destruction. Political violence has also tragically claimed the lives of many conservationists and protected area staff in Africa, Asia, and Latin America. The reemergence of civil, ethnic, and international wars and the persistence of their impacts have caused many conservationists to reassess their efforts and adapt their strategies to a new set of responsibilities and urgent challenges. These challenges include preparing conservation programs and local communities for crises; maintaining conservation capacity during periods of conflict; addressing the underlying political and economic factors that fuel war; and developing the potential of conservation to help reduce the frequency, duration, and impact of violent conflicts.

The international conference *War and Tropical Forests: New Perspectives on Conservation in Areas of Armed Conflict* assembled a diverse group of researchers, conservation practitioners, and policy makers to explore these issues and draw greater attention to the conflicts currently affecting tropical forests. The eight chapters contained in this volume emerged from their papers and presentations. Their publication follows that of other recent works addressing the role of conservation in areas of armed conflicts or the conservation implications of war (see, Westing 1993; International Union for Conservation of Nature [IUCN] 1997, 1998; Naughton-Treves 1999; Austin and Bruch 2000; Blom et al. 2000; Peluso and Watts 2001; Shambaugh, Ogelthorpe, and Ham

[Haworth co-indexing entry note]: "Preface." Price, Steven V. Co-published simultaneously in *Journal of Sustainable Forestry* (Food Product Press, an imprint of The Haworth Press, Inc.) Vol. 16, No. 3/4, 2003, pp. xvii-xxii; and: *War and Tropical Forests: Conservation in Areas of Armed Conflict* (ed: Steven V. Price) Food Products Press, an imprint of The Haworth Press, Inc., 2003, pp. xiii-xviii. Single or multiple copies of this article are available for a fee from The Haworth Document Delivery Service [1-800-HAWORTH, 9:00 a.m. - 5:00 p.m. (EST). E-mail address: getinfo@haworthpressinc.com].

2001). However, considering the proliferation of armed conflict in tropical forests areas, the amount of scholarly attention given to the relationship between war and tropical forest conservation seems disproportionately small.

War has been a widespread and persistent phenomenon in the recent history of the tropics. Armed conflict develops under a broad range of social, political, economic, and environmental conditions and its implications for forest use and management are not consistent or predictable between regions or conflicts. This volume, therefore, does not present a single view of the impact that wars have on forest conservation, nor does it reach a consensus about the role that forest resources play in the outbreak or course of armed conflict. Nonetheless, the authors collectively highlight the potential for armed conflict and military power to affect the future of tropical forests. They also draw our attention to a range of critical issues that have not customarily fallen in the purview of conservationists: the trade of small arms; the role of militaries and armed groups in the inequitable control and illicit use of forest resources; the environmental impact of refugees and internally displaced populations; and the growing social and environmental costs of international efforts to eradicate drug crops from tropical landscapes.

The response of conservationists to these diverse challenges increasingly involves collaboration with international aid organizations, governmental agencies, advocacy groups, and even militaries. This collection of essays should therefore prove useful to conservation practitioners and policy makers as well as individuals and organizations concerned with human rights, conflict resolution, rural development, international law, and foreign relations. The following paragraphs describe some of the main themes and conclusions that emerge from this volume.

The impacts of armed conflict on forest resources and conservation capacity are diverse and overwhelmingly negative–The diverse biophysical impacts of war include habitat degradation, loss of wildlife and biodiversity, increased levels of pollution, and adverse changes to human and ecosystem health. In many ways, war and insecurity can also profoundly affect the capacity of local communities, protected area agencies, and non-governmental organizations to carry out conservation activities. While the consequences of armed conflict for tropical biodiversity have been largely negative, it is important to note that in some cases hostilities have positively affected biodiversity protection and other conservation objectives. The dynamic and unpredictable nature of change during armed conflict makes it vital for conservationists to keep abreast of events and monitor the development of threats, impacts, *and* opportunities. Over the long-term, greater knowledge of the social, political, economic, and ecological dimensions of armed conflict will permit the development of more effective field strategies, policy prescriptions, and legal responses.

Conservation interests working in politically volatile regions must prepare for conflict and its aftermath–If conservation programs are to succeed–or even survive–in areas of conflict, they must be able to adapt and cope as political regimes fall, economies crumble, populations swell or shrink, and their logistical and financial support is suspended. By preparing for armed conflict and complex emergencies, conservation programs, local communities, and their partners can better cope with crises and avoid unnecessary environmental degradation. Over the past decade, the experiences of conservation organizations and protected area authorities–particularly in Africa–have yielded important lessons for improving the security and resilience of the conservation personnel and programs. These findings emphasize the critical importance of programmatic flexibility, contingency planning, political neutrality, training for junior-level staff, and collaboration with a broad range of stakeholders.

Threats to natural resource conservation can be severe during the post-conflict period–There are no simple conclusions about the relationship between war and tropical forests. However, a new maxim for conservation seems to be coined in the following pages by Jeffrey McNeely when he observes that, "while war is bad for biodiversity, peace can even be worse." In the wake of war, the regulatory authority, norms, and customs that usually govern access to and use of natural resources may be left weakened or suspended. Armed groups often take advantage of "peacetime" by facilitating illegal extractive enterprises or by directly engaging in the outright plunder of valuable natural resources. Refugees, internally displaced people, and local communities may be left more dependent on the local resource base–including protected areas and forest reserves–for food and basic supplies. If conservation programs can maintain some presence or logistical capacity during periods of armed conflict, essential conservation activities can be more promptly restarted in the crucial post-conflict period.

Local communities can play a decisive role in conservation during armed conflict–The survival or demise of forests during wartime may depend on how the needs of desperate local populations are met. One major challenge is to help local communities cope with crisis without increasing threats to the local resource base and the long-term sustainability of their livelihood strategies. This may be particularly difficult where crisis conditions aggravate existing conflicts between local interests and protected area authorities. Locally held grievances may predispose individuals to participate in unsustainable activities like wildlife poaching or logging, thus magnifying the destructive impact of war. Civil strife therefore provides another powerful argument for community-based efforts that emphasize the local benefits of conservation, strengthen community institutions, and help protect the property and customary-use rights of local people and forest-dwellers. In his contribution to this volume, David Kaimowitz observes that, "where people are willing to die to gain access to valuable natural resources, protected areas have little chance of surviving unless local groups have a strong stake in their success."

International market forces and far-reaching economic agendas often fuel armed conflict–The links that exist between armed conflict, forest resource exploitation, and powerful international market forces suggest that natural resource abundance–rather than scarcity–often fuels or perpetuates conflict. In some of the most grievous and long-lasting armed conflicts, the activities of armies, militias, and warlords are financed by the international trade of forest resources. For example, the war in the Democratic Republic of the Congo has become mainly a struggle for control over access, exploitation, and trade of the country's vast natural resources. With increasing globalization of trade and greater international consumption of tropical forest resources, the influence of market forces on armed conflict and forest resource exploitation may only grow.

Corruption and dysfunctional governance can exacerbate conflict and its environmental impact–In many countries, corrupt government agencies and military institutions facilitate the over-exploitation of natural resources, protect illicit trade networks, and reinforce inequitable patterns of resource control and use. Dysfunctional systems of governance and a lack of secure property rights are characteristic of many remote forest regions affected by armed conflict. These problems can be addressed at many different scales and in partnership with a variety of organizations and agencies that advocate human rights, arms control, certification regimes, sustainable development, and good governance.

Effective conservation in areas of armed conflict requires greater levels of collaboration in research, policy-making, and field programs–Some of the practical lessons detailed in this volume were borne of tragedies and failures in the field during wartime. Improvements to the safety, efficacy, and endurance of conservation programs can be made at a lower cost if international agencies, government ministries, relief organizations, and conservation groups share their expertise, better coordinate their agendas, and more efficiently integrate their programmatic capacities. The complexity of the challenges described in this volume also requires more resourceful and creative combinations of disciplinary approaches. The task of revealing the interrelationships between processes such as deforestation, frontier colonization, internal displacement, political violence, and international trade is daunting. Some conflicts, like Colombia's chronic and extremely complex "violence syndrome"–described in this volume by María D. Álvarez–defy most traditional explanatory frameworks. The contributors to this volume accordingly draw upon and combine historical, geographic, ecological, legal, economic, and political analyses.

The following chapters are bound together by the conviction that conservation is not a "luxury" to be pursued only in areas unaffected by war. The conflicts affecting the mega-diversity forests of countries like Colombia, Indonesia, and the Democratic Republic of the Congo present conservationists with some formidable strategic, logistical, and conceptual challenges. Such conservation crises draw our attention to the destructive impacts of violent conflict, but they also reveal the great potential for forest conservation to contribute to the survival of local communities, the prevention of conflict, and the restoration of habitats, landscapes, and economies. The grave threats associated with armed conflict warrant greater efforts to develop these promising roles for conservation. In some cases, this implies a broader conservation agenda, new responsibilities, and some unorthodox partnerships. While innovations are needed to meet these emerging challenges, the contributors to this volume remind us that the traditional principles of good conservation–including adaptive management, equitable access and control of natural resources, and international cooperation–remain essential to the future of tropical forests, in times of war and peace.

Steven V. Price

REFERENCES

Austin, J. and C. Bruch (eds.). 2000. The Environmental Consequences of War–Legal, Economic, and Scientific Perspectives. Cambridge University Press, Cambridge.

Blom, E., W. Bergmans, I. Dankelman, P. Verweij, M. Voeten, P. Wit. (eds.). 2000. Nature in War, Biodiversity Conservation During Conflicts. Mededelingen No. 37. Netherlands Commission for International Nature Protection, Amsterdam.

IUCN (International Union for Conservation of Nature–The World Conservation Union) (ed.). 1997. Parks for Peace Conference Proceedings. IUCN, Gland, Switzerland.

IUCN. 1998. International Symposium on Parks for Peace, Stelvio National Park, Bormio, Italy, 17-21, May 1998. Conference Proceedings. IUCN, Gland.

Naughton-Treves, L. (ed.). 1999. Fighting in the Forest, Biodiversity Conservation Amidst Violent Conflict. CDF Discussion Paper. Conservation and Development Forum, Gainesville.

Peluso, N.L. and M. Watts (eds.). 2001. Violent Environments. Cornell University Press, Ithaca.

Shambaugh, J., J. Ogelthorpe, and R. Ham (with contributions from Sylvia Tognetti). 2001. The Trampled Grass: Mitigating the impacts of armed conflict on the environment. Biodiversity Support Program, Washington, D.C.

Westing, A.H. (ed.) 1993. Transfrontier Reserves for Peace and Nature: A Contribution to Human Security. United Nations Environmental Programme (UNEP), Nairobi.

Acknowledgements

This volume had its genesis in the international conference *War and Tropical Forests: New Perspectives on Conservation in Areas of Armed Conflict*. Organized by graduate students working under the auspices of the Yale Student Chapter of the International Society of Tropical Foresters (Yale ISTF), the conference was held at the Yale School of Forestry and Environmental Studies on March 31 and April 1, 2000. The essays in this volume were originally presented in some form at that meeting. Over the past year, the authors have updated their material, and in some cases, they have thoroughly rewritten their essays. Their participation in the conference and their contributions to this volume are deeply appreciated. A debt of gratitude is also due to the large group of expert reviewers who provided detailed and invaluable comments on earlier drafts of this volume. Three papers presented at the conference are not included here, but I wish to thank Patrick Alley of Global Witness, Jamison Suter of Fauna and Flora International, and independent author and journalist Bill Weinberg for their invaluable contributions to the conference.[1]

Special thanks are owed to the series editor of the *Journal of Sustainable Forestry*, Professor Graeme P. Berlyn, for proposing this publication and for providing continuous logistical support. His assistant, Uromi Goodale, also helped facilitate the preparation of this publication. From the inception of the conference, Professor Mark Ashton provided invaluable guidance as Yale ISTF's stalwart faculty advisor and director of Yale's Tropical Resource Institute. It is also important to acknowledge Jay Austin and Carl Bruch at Environmental Law Institute, Thomas Dillon at World Wide Fund for Nature (WWF), and Jefferson Hall at Wildlife Conservation Society (WCS) for providing many invaluable ideas and contacts that helped shape the conference and therefore, this volume. Professor David Watts at the Anthropology Department of Yale University generously provided expert guidance on the editing of sections related to gorilla conservation. He and fellow professors Arun Agrawal, Enrique Mayer, James Scott, Daniela Spenser, and Eric Worby served as moderators during the conference and greatly enhanced the panel discussions. A special note of appreciation is also due to Arthur W. Galston, Eaton Professor Emeritus of Molecular, Cellular, and Development Biology at Yale University, who provided inspirational closing remarks to the conference. Philip R. S. Johnson also provided suggestions that helped improve the preface to this volume.

The conference and this publication would not have been possible without the financial support provided by many programs and offices at Yale University. Thanks are therefore due to: Dr. Jim Bryan and the Tropical Resource Institute; Dean Gus Speth of the Yale School of Forestry and Environmental Studies; the Graduate School of Arts and Sciences symposium fund; Professor Donald Green and the Institution for Social and Policy Studies; the Yale Council on Latin American Studies; and the Class of 1980 Fund of the Yale School of Forestry and Environmental Studies.

NOTE

1. Bill Weinberg's conference paper was drawn from his comprehensive and uncompromising book *Homage to Chiapas, The New Indigenous Struggles in Mexico* (2000, Verso: New York). Patrick Alley's paper examined war, corruption, and forest policy in Cambodia, and drew upon Global Witness' authoritative reports on Cambodia. Global Witness' reports on war-related environmental issues in Asia and Africa are available at: http://www.globalwitness.org

Biodiversity, War, and Tropical Forests

Jeffrey A. McNeely

SUMMARY. Tropical forests are one of the world's last remaining frontiers. Like all frontiers, they are sites of dynamic social, ecological, political, and economic changes. Such dynamism involves constantly changing advantages and disadvantages to different groups of people, which not surprisingly leads to armed conflict, and all too frequently to war. Many governments have contributed to conflict by nationalizing their forests, so that traditional forest inhabitants have been disenfranchised while national governments sell trees to concessionaires to earn foreign exchange. Biodiversity-rich tropical forests in Papua New Guinea, Indonesia, Indochina, Myanmar, Sri Lanka, Central Africa, the Amazon, Colombia, Central America, and New Caledonia have all been the sites of armed conflict, sometimes involving international forces. While these conflicts have frequently, even invariably, caused negative impacts on biodiversity, peace is often even worse, as it enables forest exploitation to operate with impunity. Because many of the remaining tropical forests are along international borders, international cooperation is required for their conservation; as a response, the concept of international "peace parks" is being promoted in many parts of the world as a way of linking biodiversity conservation with national security. The Convention on Biological Diversity, which entered into force at the end of 1993 and now

---

Jeffrey A. McNeely is Chief Scientist, IUCN–The World Conservation Union, Biodiversity Policy Coordination Division, Rue Mauverney 28, 1196 Gland, Switzerland (E-mail: jam@iucn.org).

The original version of this paper was presented at the conference "War and Tropical Forests: New Perspectives on Conservation in Areas of Armed Conflict," held on March 31 and April 1, 2000, Yale School of Forestry and Environmental Studies.

[Haworth co-indexing entry note]: "Biodiversity, War, and Tropical Forests." McNeely, Jeffrey A. Co-published simultaneously in *Journal of Sustainable Forestry* (Food Product Press, an imprint of The Haworth Press, Inc.) Vol. 16, No. 3/4, 2003, pp. 1-20; and: *War and Tropical Forests: Conservation in Areas of Armed Conflict* (ed: Steven V. Price) Food Products Press, an imprint of The Haworth Press, Inc., 2003, pp. 1-20. Single or multiple copies of this article are available for a fee from The Haworth Document Delivery Service [1-800-HAWORTH, 9:00 a.m. - 5:00 p.m. (EST). E-mail address: getinfo@haworthpressinc.com].

has nearly 180 State Parties, offers a useful framework for such cooperation. *[Article copies available for a fee from The Haworth Document Delivery Service: 1-800-HAWORTH. E-mail address: <getinfo@haworthpressinc.com> Website: <http://www.HaworthPress.com>*

KEYWORDS. Tropical forest, large mammals, biodiversity, war, indigenous people, frontiers

INTRODUCTION

The "peace dividend" expected from the end of the Cold War has not paid off in terms of reduced violent conflict, and the recent nuclear weapons tests by India and Pakistan demonstrate the continuing potential for highly destructive war. Some tropical countries are facing generalized lawlessness and banditry, including by marauding ex-soldiers in several African nations and drug cartels in some parts of Latin America (Renner 1996). Tension in various parts of Africa, Central America, Indonesia, Colombia, Sri Lanka, and elsewhere are further indications of war as a fact of modern life in tropical many forest countries.

Despite these widespread threats to national sovereignty, governments are obliged under the 1992 Convention on Biological Diversity to conserve their own biodiversity (Article 1) and to ensure that activities within their jurisdiction or control do not cause damage to the environment of other states (Article 3). Any negative impacts of war on biodiversity clearly are contrary to this international agreement. But what, specifically, are the impacts of war on biodiversity in tropical forest countries? This paper attempts to identify some of the key issues in preparing a balanced assessment.

The issues are complicated and the available evidence does not provide simple answers. However, it is hard to avoid the conclusion that various trends will make future wars extremely destructive for both people and the rest of nature. Some of the most influential factors include modern means of communication, growing human populations and levels of resource consumption, increased vulnerabilities of interdependent, integrated civil societies, and the spread of modern instruments of war.

On the other hand, war is often seen as part of the way human societies adapt to changing conditions (see, for example, Harris 1974; Keeley 1996; Vayda 1974). The International Commission on Peace and Food

(1994) concluded that historically, all the major changes in the international political and security system have been the result of armed conflicts, wars, and revolutions. It appears that many, even most, societies have been defined by war, and that the organization of a society for the possibility of war has been its principal political stabilizer. The victors who emerged from the ashes of war have sown the seeds that would produce subsequent tensions, disputes, and conflicts. It often seems that an institutional lack of capacity to adapt to change, or the inertia of vested interests in the *status quo*, means that societies inevitably become maladapted, eventually requiring a shock such as war to set them on a different course (Edgerton 1992).

A fundamental issue is how humans stay within the productive limits of their supporting ecosystem. While most would agree that such adaptation should be possible through the application of knowledge and wisdom, history does not support such a rational view, and in fact war is virtually universal in human societies as a means of resolving conflicts arising from various sources of maladaptation (Keeley 1996). Underlying stress factors can produce or deepen rifts in societies, with disputes triggered by glaring social and economic disparities and exacerbated by the growing pressures of resource depletion, natural calamities, environmental degradation, and perceived excess population. Biodiversity-related problems like desertification, soil erosion, deforestation, and water scarcity reduce food-growing potential, worsen health effects, and diminish life-support capacity, which can lead to civil conflict and increase the likelihood of war. Based on experience in Nicaragua, Nietschmann (1990a, 37) concludes, "Degraded land and resources are as much a reason for taking up arms as are repression, invasion, and ideology."

Because environmental stress can be a fundamental cause of armed conflict, issues of biodiversity conservation, the sustainable use of biological resources, and the fair and equitable sharing of the benefits of such uses–the three objectives of the Convention on Biological Diversity–are critical elements in discussions of national security in tropical forest countries. Investments in such activities as sustainable forestry, water conservation, land reform, and protected areas management, it can be argued, are vital contributions to peace.

This paper will begin by briefly assessing war as one of the traditional social means for human societies to adapt to changing environmental conditions, then assess some of the positive and negative impacts of war on tropical forest biodiversity. It then suggests several issues that must be addressed if modern civilization is to meet the growing security challenges of the 21st century. It will conclude by showing how conserving biodiversity can contribute to peace, building on the preamble to the Convention on Biological Diversity, which states that, "Ultimately, the conservation and sustainable use of biological diversity will strengthen friendly relations among states and contribute to peace for humankind."

THE HISTORY OF WAR AND BIODIVERSITY

Today's biodiversity is to a considerable extent the result of long-term interactions between people and their environments reaching back at least as far as the origins of fire (see, for example, Flannery 1994; McNeely 1994; Martin and Klein 1984; Ponting 1992). The greatest diversity of terrestrial species today is found in forested areas inhabited by tribal and other indigenous peoples, where relatively large areas of "unoccupied" territory serve as a sort of buffer zone between communities that may be embroiled–at least historically–in virtually constant warfare, including sneak attacks, revenge killings, kidnappings, and raids on livestock (Keeley 1996). It is instructive, therefore, to briefly examine the impact on biodiversity of warfare among traditional and indigenous societies, how modern armies relate to tropical forest-dwelling tribal peoples, and the influence such relations have had on biodiversity.

Ember and Ember (1992) found that higher frequencies of war in traditional societies can be forecast by a history of unpredictable natural disasters and severe food shortages, as people have tried to protect themselves by going to war to take resources from enemies. Raids often included plundering food stores and gardens in the Americas, Polynesia, New Guinea, and Africa, leaving an enemy facing starvation and rendering large areas of territory at least temporarily uninhabitable. While this could serve to provide larger areas of habitat to various species of wildlife, it could also lead to significant increases in the pressure of human population on the remaining wildlife populations. Losses and gains of territory were a very frequent result of warfare among pre-industrial societies, leading to dynamic tribal boundaries; and these frontiers often were places supporting great diversity of species. Keeley (1996, 112) concludes, "Even in situations where no territory exchanges hands, active hostilities along a border can lead to development of a no-man's-land, as settlements nearest an enemy move or disperse to escape the effects of persistent raiding. Although such buffer zones could function ecologically as game and timber preserves, they were risky to use even for hunting and wood cutting because small isolated parties or individuals could easily be ambushed in them."

These buffer zones often are where biodiversity is richest, especially in terms of large mammals. Consider as just one example, South America at the time of the first contact with Europeans. Large settled villages were found along the major rivers in various parts of the Amazon. The chieftains of these societies practiced a type of warfare that often involved forces numbering in the hundreds of men drawn from multiple confederated villages who traveled by canoes and used sophisticated tactics to attack their enemies. The powerful chieftains often fought over territory, with large buffer zones separating them; these buffer zones often were refugia for wild game (Ferguson 1989b). In the first voyage up the Amazon's Ucayali River in 1577, Juan Salenas Deloyola contacted three principal groups, similar in culture but speaking different languages (an indication of linguistic separation). A distance of 50 to 60 leagues separated each group from the next, about the same distance as was incorporated in the tribal territory. Myers (1979) considers this an example of a no-man's land, located between the defended territories of adjacent human groups. While the evidence available at present does not support any particular conclusions about the relationship between ecology and war, competition for environmental resources very frequently is a factor in war between different communities in Amazonia (Ferguson 1989a). Vulnerability to attack may set a threshold on settlement size, or the threat of raids may encourage people to live together to maintain an adequate defensive force.

Warfare between modern and traditional societies has often involved what might be termed "ecological attacks." As one of the most obvious examples of this, the final destruction of the great herds of American bison (*Bison bison*), the foundation of Plains Indian life in North America, closely coincided with the defeat of the Sioux and Cheyenne in the 1870s. Biological warfare was also used, either accidentally or intentionally. Diseases such as smallpox, measles, and influenza had a major impact on the native populations of the Americas, Australia, and the Pacific Islands because they lacked immunity to the "new" diseases. Perhaps more important, the Europeans also brought ecological transformations that disrupted traditional economies and replaced native ecosystems with new agricultural systems that produced more of the goods required by colonists (Crosby 1986), leading to fundamental–and perhaps permanent–changes in biodiversity.

One of the world's biologically richest areas is in the upper Amazon, including Venezuela, Colombia, and Brazil: a true "biodiversity hotspot" (McNeely et al. 1990), where borders are not well demarcated. Perhaps not coincidentally, this is also an area that is occupied by a large number of culturally distinct Indian groups that have formed long-term relationships with their environment. These relationships included elements such as warfare, infanticide, and raiding, that are unacceptable in modern society (except, of course, where they are sanctioned by the government as part of modern warfare). For example, Chagnon (1988) has found that among the Yanomamo Indians, the largest Indian group in the Amazon rainforest, 44% of males 25 or older have participated in the killing of someone, about 30% of adult male deaths are due to violence, and nearly 70% of all adults over 40 have lost a close genetic relative due to violence. The relationship between indigenous peoples, biodiversity, colonists, and the modern military in this frontier region is a complex and fascinating one that contains several important lessons for those seeking better understanding of the relationship between biodiversity and national security in tropical forest countries.

In November 1981, Brazil's President Fernando Color de Melo issued a decree to give the Yanomamo partial control of their traditional lands. The decree was opposed by the Brazilian military, because the Yanomamo lands extend across the borders with Venezuela and Colombia, a militarily sensitive area. The decree was part of a zoning process which involved dividing the forest into protected areas, land for traditional Indian farming and hunting, and areas permitting environmentally destructive development such as logging, roads, mines, and dams.

However, the Brazilian military has continued to impede full legalization of Indian land rights near its international borders, branding as subversives those scientists who are working internationally to save the Amazonian forest habitats of the indigenous peoples. Lewis (1990) reported on a secret document prepared by the Brazilian High War College proposing that war could be used against indigenous or environmental organizations in the Amazon. The idea that the Amazon might be invaded by a foreign army of conservationists aiming to conserve the rainforest may appear ludicrous to those living outside South America, but it is taken seriously in the region and has been used to justify the Brazilian military's tight control of Amazonian policy (Conklin and Graham 1995).

Conselho Indigenista Mission (CIMI) concludes that the Brazilian military sees the preservation of the rainforest and its peoples as a threat to national security (CIMI 1987), considering it necessary to "clean" the frontier strip of obstacles to the implantation of more permanent investments, which spells disaster for the Indians and for biodiversity. This perception perpetuates the conflict among the military, indigenous peoples, and conservation interests.

This military mind-set is not confined to Brazil. In Venezuela, a proposal to create a Yanomamo Biosphere Reserve along the border with Brazil was rejected by the Ministry of External Relations, concerned that national and international public opinion would be mobilized to seek the human rights of the indigenous groups and to promote eventual self-development and self-determination. The ministry singled out a group of Venezuelan ecologists and anthropologists as the core of an international conspiracy to undermine the ability of the government to control the Amazon territory and its native inhabitants (Hill 1994). The high-level Congress of the Armies of the Americas (CAA) used a number of distortions to reduce complex social problems into a black and white opposition between "national security" and "terrorist subversion," with those advocating Indian rights being linked to subversive organizations (a group that also included feminists and environmentalists). In essence, the CAA created a mythological history of the relationships between indigenous peoples and their land. It defined problems in terms that required military solutions (Hill 1994), ignoring the role of indigenous ways of life in maintaining the rich biodiversity of the upper Amazon, and their dependence of the forest-dwelling people on the biological resources of the forest.

New Guinea is a tropical forest-covered island that has been a particularly fertile ground for the study of war, as warfare has been frequent, deadly, and a defining factor in the life of most tribal peoples of the island during the time anthropologists were available to study its highly diverse societies (over 700 languages are known from New Guinea). For example, warfare among the Maring, a people of the New Guinea Highlands, facilitated demographic shifts, adjusted relationships between population and land, and alternated the build up of pig herds with slaughter for pig feasts that played an important role in warfare. Rappaport (1984) saw warfare as part of a self-regulating ecological system that maintained the population of both people and pigs below the carrying capacity of the land. Some of the New Guinea highland cultures have particularly bloody histories. For example, the Mae Enga fought 41 wars for land between 1900 and 1950, of which six resulted in complete routs of the enemy that led to acquisition of new territory from the defeated clan (Meggitt 1977). Among the Dani people of the New Guinea

Highlands, warfare is responsible for almost 30% of mortality (Heider 1970). Warfare in association with hunting has been well documented among a number of other New Guinea groups, including the Purari, the Kiwai, the Trans-Fly peoples, the Marind-Anim, the Kolopom, Jacquia, and the Asmat.

Generally speaking, the New Guinea tribes engage in two rather different kinds of warfare. One is highly ritualistic, involving hundreds of men who meet in a designated public battleground and shoot arrows at each other; these battles tend to be generally inclusive and casualties are low. The other kind of warfare is more secular, brief, and infrequent. It often involves a large-scale clandestine attack that kills large numbers of people and destroys property (Shankman 1991). Some battles lead to massacres of over 100 people in an hour or so (Blick 1988), which can amount to over 5% of the group's population (an impact equivalent to 14 million Americans dying). Heider (1979) sees New Guinea warfare as a cycle of battles and raids over many years that constantly splits alliances and rearranges confederations, thus setting the stage for subsequent battles. The result of such fighting is that fields and home sites are abandoned, thereby redistributing land and other resources, and creating buffer zones that provide sanctuary to at least some components of biodiversity.

Indigenous warfare was prevalent throughout Melanesia, and anthropological accounts of pre-colonial warfare come from the Admiralty Islands, New Ireland, New Britain, Bougainville, Choiseul Island, New Georgia, Malaita, San Cristoval, New Hebrides (now Vanuatu) and New Caledonia, and both coastal and interior New Guinea (summarized in Knauft 1990).

While the existence or intensity of warfare in pre-state societies is not a simple linear function of population density, population pressure, or protein scarcity, all of these factors are important contributors. It seems reasonable to conclude that ecological pressure works together with cultural and political dispositions toward warfare. The perception of individual or group land scarcity is a function of sociocultural as well as ecological organization, and perceptions of scarcity are often as important as the pattern of rainfall, the numbers of pigs, or the game animals in the forest (Knauft 1990). Thus, the actual warfare carried out by the indigenous peoples of the tropical forests involved numerous factors reinforcing each other, including increasing human population density, related clearance of forests to increase domestic food production, and declining wild food resources. The concomitant rise in the demand for resources led to increased opportunities for conflict. It is certain that the subsequent population redistribution had implications for biodiversity.

To conclude this section, it appears that various forms of war have been part of the way traditional societies adapted to changing conditions, and–at least coincidentally–helped contribute to the rich biodiversity found today in many tropical forest areas occupied by traditional and indigenous peoples. Bringing peace to these regions will remove this means of adaptation, requiring other ways to conserve biodiversity and maintain the capacity to adapt to changing conditions.

THE IMPACTS OF WAR ON BIODIVERSITY IN TROPICAL FORESTS

Negative Impacts of War on Biodiversity

The negative impacts of war on biodiversity in tropical forests result from the collective actions of large numbers of people for whom war is a dispensation to ignore normal restraints on activities that cause environmental damage. War, and preparations for it, have negative impacts on all levels of biodiversity, from genes to ecosystems. These impacts can be direct–such as hunting and habitat destruction by armies–or indirect, for example through the activities of refugees.

Sometimes these impacts can be deliberate, and a new word has been added to the military vocabulary: "ecocide," the destruction of the environment for military purposes; it clearly builds on the "scorched earth" approach of earlier times. Westing (1976) divides deliberate environmental manipulations during wartime into two broad categories: those involving massive and extended applications of disruptive techniques to deny to the enemy any habitats that produce food, refuge, cover, training grounds, and staging areas for attacks; and those involving relatively small disruptive actions that in turn release large amounts of "dangerous forces" or become self-generating. An example of the latter is the release of exotic micro-organisms or spreading of landmines (of which over 100 million now litter active and former war zones around the world) (Strada 1996).

This discussion could be long and dreary, but only a few illustrative cases will be mentioned. Perhaps the most outstanding example is Vietnam, where US forces cleared 325,000 ha of land and sprayed 72,400 m^3 of herbicides in the name of security (Westing 1982). The impact on biodiversity was severe; spreading herbicides on 10% of the country (including 50% of the mangroves) led to extensive low-diversity grasslands replacing high-diversity forests, mudflats instead of highly productive mangroves, major declines in both freshwater and coastal fisheries, and so forth (Nietschmann 1990a).

Other problems are more systemic. The State Law and Order Restoration Council (SLORC), the military government in Myanmar (formerly Burma), has been involved in violent confrontations with many of the tribal groups who inhabit the densely forested mountain regions along the country's borders with Bangladesh, India, China, Laos, and Thailand. Some of these tribal groups, such as the Karen, have turned to intensive logging to fund their war effort, even though such over-exploitation will eventually destroy the forest cover and make them more open to attack (Harbinson 1992). The general lawlessness along the Thai border has greatly increased the flow of logs, both with and without government permission, leading to the virtual clear felling of many of the country's most productive forests.

Africa provides several recent war-related disasters for biodiversity in tropical forests. Like the upper Amazon, the Virunga Volcanoes region (including parts of the Central African countries of Rwanda, Democratic Republic of the Congo, and Uganda) is exceptionally rich in species diversity, including the rare and endangered mountain gorilla (*Gorilla gorilla beringei*) whose total population is approximately 600. The civil war against the government of Rwanda was launched in 1990 from within the Virunga Volcanoes region, spreading deeper into Rwanda until 1994 and sending large numbers of refugees fleeing to North Kivu District in what was then Zaire, which then began a civil war of its own. The headquarters of several tropical forest World Heritage sites in Zaire were taken over by the military, including Virunga National Park, Kahuzi-Biega National Park and the Okapi Wildlife Reserve. In 1994, some 850,000 refugees were living around Virunga National Park, partly or completely deforesting some 300 km^2 of the park in a desperate search for food and firewood. Up to 40,000 people entered the park every day, taking out between 410 and 770 tons of forest products. The bamboo forests have been especially seriously damaged, and the populations of elephants, buffalo, and hippos have been much reduced. Organizations such as the Red Cross, Médecins Sans Frontière, and CARE have supported well-meaning relief operations on the park boundaries and have even established a dump for medical wastes inside the park, with obvious disease transmission risks associated with such practices (Pearce 1994). At least 80 of Virunga's park staff have been killed in battle with insurgents since 1996.

A few other examples (among many that could be provided):

- The administrator and two rangers of the Saslaya National Park in Nicaragua (15,000 ha) were kidnapped by the Contras in 1983, forcing the National Environment Agency to abandon the management of the area (Thorsell 1990).
- In 1996, the Kibira and Ruvubu National Parks in Burundi were used as sanctuaries and entry points for guerrillas fighting the government. As a result, they also became operational areas for government troops, with both sides heavily involved in poaching (Winter 1997).
- India's Manas Wildlife Sanctuary, a World Heritage site, has been taken over by guerrillas from the Bodo tribe, who have burned down park buildings, looted most park facilities, killed guards, destroyed bridges, poached rhinos (*Rhinoceros unicornis*), elephants (*Elephas maximus*), tigers (*Panthera tigris*), and other wildlife, cleared forest and depleted fish stocks in the Manas River.
- In Sri Lanka, Tamil rebels attacked Wilpattu National Park in 1989, killing over a dozen guards and destroying facilities. This caused a withdrawal of conservation staff and a great increase in military activity.
- Liberia's civil war has forced rural people to hunt duikers (*Cephalophus* spp.), pygmy hippos (*Choeropsis liberiensis*), forest elephants (*Loxodonta*), and chimpanzees (*Pan troglodytes*) for food (Wolkomir and Wolkomir 1992).
- During the Vietnam War, elephants were specifically targeted by helicopter gunships because they might be used as pack animals by the Viet Cong. In Sudan, the white rhino (*Ceratotherium simum*) was exterminated during the 17 years of civil war from 1955 to 1972 (Abdullah 1997).
- In the Democratic Republic of the Congo, civil war has stopped efforts to protect the last habitat of the pygmy chimpanzee, or bonobo (*Pan paniscus*), a species endemic to that country. Fewer than 15,000 of the apes survive, and they are increasingly threatened by local people who are being forced to depend on the forest for survival. This includes hunting of bonobos for bushmeat, which many westerners consider just one small step removed from cannibalism. One western researcher reported that poachers and army deserters armed with machineguns are hunting in Selonga National Park, a World Heritage site that is a stronghold of this species.

The conclusion is not surprising: war is bad for biodiversity.

Positive Impacts of War on Biodiversity

However, war, or the threat of war, can also be good for biodiversity, at least under certain conditions. As Myers (1979, 24) put it, "In some respects, indeed, wildlife benefits from warfare: combatant armies effectively designate war zones as 'off limits' to casual wanderers, thus quarantining large areas of Africa from hunters and poachers." Of course, any benefits of war to biodiversity are incidental, inadvertent, and accidental rather than a planned side effect of conflict. But even so, it is useful to review some cases where war, or preparations for war, has benefited biodiversity, perhaps supporting the views of some anthropologists that war helps societies adapt to their environmental constraints.

For example, the border between Thailand and Peninsular Malaysia was a hotbed of insurgency during the mid-1960s to mid-1970s. On the Malaysian side of the border, the military closed off all public access and potential logging activity in the Belum Forest Reserve. As a result, this extensive area of some 160,000 ha has remained untouched by modern logging pressures and therefore is rich in wildlife resources. Malaysia is now converting this into a national park that will form a transboundary protected area with matching protected areas in southern Thailand.

While the second Vietnam War was an ecological disaster, it also led to some important biological research, such as the extensive, long-term, review of migratory birds in eastern Asia carried out by the Migratory Animals Pathological Survey (McClure 1974). The excuse for this research was its relevance to the war effort, but it has yielded data that are useful for numerous civilian conservation applications. The watersheds through which ran the Ho Chi Minh Trail, some of the most heavily bombed parts of Indochina during the second Vietnam War, have more recently been remarkably productive in new discoveries. The new large mammal discoveries include two species of muntjak or barking deer (*Megamuntiacus vuquangensis* and *Muntiacus truongsonensis*), a unique variety of forest antelope (*Pseudoryx nghetinhensis*), and a bovid ultimately related to wild cattle (*Pseudonovibos spiralis*) (Dillon and Wikramanyake 1997) as well as the rediscovery of a species of pig that formerly was known only by a few fragmentary specimens. That such species could survive in such a heavily bombed area is testimony to the recuperative power of nature, and the ability of wildlife to withstand even the most extreme kinds of human pressure during warfare. Interestingly, these species now are even more severely threatened by the peacetime activities of development than they were by the Indochina wars.

Some other species are likely to have benefited from the war in Vietnam. Orians and Pfeiffer (1970, 553) say that tigers "have learned to associate the sounds of gunfire with the presence of dead and wounded human-beings in the vicinity. As a result, tigers rapidly move toward gunfire and apparently consume large numbers of battle casualties. Although there are no accurate statistics on the tiger populations past or present, it is likely that the tiger population has increased much as the wolf population in Poland increased during World War II."

Fairhead and Leach (1995) report that parts of the Ziama region of Guinea, which includes an extensive biosphere reserve, became forested following a series of wars that affected the area from 1870 to 1910. The resident Toma people first fought with Mandinka groups from the north and subsequently with the French colonial armies, causing major depopulation and economic devastation that in turn allowed the forest to reclaim agricultural land. The human disaster of war enabled nature to recover.

The impact of war on biodiversity is often decidedly mixed, with a complex combination of damages and benefits (see Table 1). Nicaragua provides an outstanding example. Engaged in civil war for over 20 years, the country suffered 100,000 casualties, and nearly half of its population was relocated in one way or another. The human tragedy was immense, but biodiversity was able to recover from a long history of exploitation, as trade in timber, fish, minerals, and wildlife was sharply reduced. The domestic cattle population, which was roughly equivalent to the human population when the war started, was reduced

TABLE 1. Impacts of War on Biodiversity

Negative Impacts	Positive Impacts
• Deforestation • Erosion • Wildlife poaching • Habitat destruction • Pollution of land and water • Reduces funds for conservation • Stops conservation projects • Forces people onto marginal lands • Creates refugees who destroy biodiversity	• Creates "no-go" zones • Slows or stops developments that lead to loss of biodiversity • Focuses state resolve • Reduces pressure on some habitats • Allows vegetation to recover in some areas • Disarms rural populations, thereby reducing hunting • Can increase biodiversity-related research

by two-thirds, freeing pastures for re-colonization by forests. This enabled the recovery of animal populations such as white-tailed deer (*Odocoileus virginianus*), peccaries (*Tayassu angulatus*), four species of monkeys (family Cebidae), caiman (*Caiman crocodilus*), iguanas (*Iguana iguana*), large birds, and various mammalian predators. Fishing boats were destroyed and fishermen fled, leading to drastic declines in the catches of fish, shrimp and lobsters, which in turn revitalized these fisheries. On the other hand, some hunting by soldiers had at least local negative impacts on wildlife, and new military bases and roads were established in formerly remote areas, opening them up to exploitation. Further, the country's once outstanding system of protected areas fell into neglect, and new areas planned were not established; the collapsing economy forced villagers into environmentally destructive activities, including clearing forest for firewood and harvesting wildlife for food. Nietschmann (1990b) concludes that a significant portion of this conflict was over resources and territory, not ideology. Biodiversity rejuvenated by the war came under renewed threat by people whom the war had impoverished; the post-war period saw a great acceleration of such impacts and now that peace has broken out, biodiversity is under renewed pressure.

On the other side of the world, the war in Indochina was disastrous to Cambodia, in both human and ecosystem terms. Years of fighting have created a climate of lawlessness in which those who control the guns also control the country's most valuable natural resources, namely forests and fisheries. Overturning any feeble efforts at control, both are being depleted at dangerous rates, according to studies being carried out by the World Bank and the Asian Development Bank (ADB). Uncontrolled logging, much of it illegal, could virtually deforest the country within five years, according to ADB, with current harvesting over three times the sustainable yield. The fish, especially from Cambodia's Tonle Sap (Great Lake), are being over-harvested, primarily for export to surrounding–and wealthier–countries. The ecological productivity of the lake was based largely on the 10,000 km^2 of flooded forest that ensured a healthy flow of nutrients into the lake. However, less than 40% of the flood forest remains under natural vegetation. Since 1993, military commanders have come to regard the forest resources as their own resources, treating them as a supplemental source of finance irrespective of the long-term impact on the country's security. Continuing loss of forests will further affect the climate, cause erosion that fills irrigation channels and fishing grounds with silt, and leave Cambodian farmland more vulnerable to both drought and flooding. This complex of problems is very similar to that which faced Cambodia some 400 years ago, when the great civilization centered on Ankor Wat collapsed under environmental pressure (McNeely and Wachtel 1988).

So, while war is bad for biodiversity, peace can be even worse. In the 1960s, when Indonesia and Malaysia were fighting over border claims on the island of Borneo, they did relatively little damage to its vast wilderness, but in the 1990s, they peacefully competed to cut down and sell its forests. In Indonesia, the 1997-98 forest fires that caused US$4.4 billion in damage were set primarily by businesses and military to clear forests in order to plant various cash crops. Ironically, the prices of these commodities that were to be grown have fallen considerably in recent years, making them even less profitable. Vietnam's forests are under greater pressure now that peace has arrived than they ever were during the country's wars; and Nicaragua's forests are now under renewed development pressures. Laos is paying at least part of its war debts to China and Vietnam with timber concessions. I was told in Laos that the Chinese and Vietnamese timber merchants and logging companies are able to operate with impunity there, irrespective of logging regulations, protected area boundaries, or any other considerations. This is perhaps not surprising given the dependence of the Pathet Lao on the support of Vietnam and China during the Indochina wars. The motivations may be more noble in times of peace, but the impacts of inappropriate development on biodiversity often are even worse than the impacts of war. Market forces may be more destructive than military forces.

CONCLUSIONS

National and international security can no longer be conceived in narrow military terms. Ethnic conflict, environmental degradation, pollution, and famine leading to civil unrest or massive migrations of refugees constitute threats to both social stability and the preservation of a productive material base–the planet's biodiversity. Thus stopping deforestation or augmenting food production capabilities in deficit areas directly and substantially contribute to the security of society, and can help prevent–or at least postpone–armed conflict. Allocating international resources to environmental monitoring and impact assessment, protection of economically important species, quick response to disasters and accidents, energy conservation, and the minimization and management of waste are all highly appropriate activities that will prevent strife and therefore reduce the likelihood of conflicts leading to war. As Peter Thacher (1984, 12) put it, "Trees now or tanks later."

More broadly, some countries are recognizing the possibility of using protected areas designed to conserve biodiversity along their borders as ways of promoting peace (e.g., Hanks 1998). In many countries, boundaries are found in mountainous areas that also tend to be biologically rich because of the great variety of habitats and ecosystem types found within relatively small areas, affected by differences in elevation, microclimate, and geological factors. While such ecologically diverse areas are often particularly important for conservation of biodiversity, they also are frequently sanctuaries in war, especially civil wars and guerrilla wars.

Peace Parks are far more than a fond hope. Peru and Ecuador fought three territorial wars in the 20th century, but Peruvian president Alberto Fujimori and Ecuadorian president Jamil Mahuad resolved their violent border dispute in 1998 with an innovative plan that included creation of two national "peace parks" near the most contested stretch of their frontier. Four mediators–the United States, Argentina, Brazil, and Chile–helped resolve the hottest regional dispute in South America through binding arbitration. The agreement also granted Ecuador free trade and navigational access to the economically important shipping routes of Peru's Amazon territory. While the agreement fell far short of Ecuador's desire for sovereignty over the disputed territory, leading to demonstrations against the government, many of Ecuador's economic goals were achieved. The area is also the territory of several Jivaro-speaking tribes, who frequently are at war with each other. The new peace with protected areas will need to involve the indigenous peoples as well (Faiola 1998).

Given that national frontiers are especially sensitive areas where conflict is endemic and biological resources are especially rich, the idea of establishing protected areas on both sides of the border–as so-called "peace parks"–has attracted considerable attention way (see, for example, Westing 1993; Westing 1998; and Thorsell 1990). It has provided a symbol of the desire of the bordering countries to deal with many of their problems in a peaceful manner. Zbicz and Greene (1998) have found that transboundary protected areas cover well over 1.1 million km^2, representing nearly 10% of the total area protected in the world (see Table 2). In addition to indicating the importance of transboundary protected areas, this also demonstrates how much of the world's land area devoted to biodiversity conservation is in remote frontier areas where risks of war historically are highest. Brock (1991) concludes that

TABLE 2. Transfrontier Protected Areas in Tropical Forest Regions

Many protected areas are located on national borders, and some have adjacent protected areas on the other side of the border, forming complexes that could be the focus of collaboration. IUCN (1997) calls these (perhaps optimistically) "Parks for Peace." The following is an indication of how widespread and important such areas are.

Continent	Transfrontier Protected Area Complexes	Designated Protected Areas
Africa	39	110
Asia	31	74
Latin America	35	89
TOTALS:	**105**	**273**

Compiled on the basis of information presented in IUCN (1997)

although peace parks have probably had relatively little independent effect on international relations, transfrontier cooperation on biodiversity issues has the potential to develop into an important factor in regional politics. Peace parks can help to internalize norms, establish regional identities and interests, operationalize routine international communication, and reduce the likelihood of the use of force.

Such areas also need to be ready to adapt to unstable conditions. Hart and Hart (1997, 309) conclude that "the best preparation for conservation in the face of regional instability is the professional development of national staff and strong site-based conservation programmes." But a key element is that these site-based initiatives must be tied to an international structure that endures when nations crumble. They propose establishing a fund that provides for continued professional development and support for field activities by the staff of protected areas during crisis periods. Such support might be focused on specific sites of international biological significance with the goal of developing semi-autonomous management within those areas. The mission of the proposed fund would be to build professional identity in national staff where national institutions have failed and to facilitate their reintegration into conservation activities after the crisis has passed.

To conclude, trying to tease out causality in the relationship between war and biodiversity issues in tropical forests is highly complex, because individuals make multiple, mutually constraining decisions that are shaped by interacting environmental and social conditions, all of which have themselves multiple interrelationships. People often learn through conflict, as fundamental interests are challenged. As Lee (1993,

10) points out, "Conflict is necessary to detect error and to force corrections. But unbounded conflict destroys the long-term cooperation that is essential to sustainability. Finding a workable degree of bounded conflict is possible only in societies open enough to have political competition."

REFERENCES

Abdulla, R. 1997. Protected areas during and after conflict. Nimule National Park: a case study. Pp. 195-199 in IUCN (ed.). Parks for Peace Conference Proceedings. IUCN, Gland, Switzerland.

Blick, J.P. 1988. Genocidal warfare in tribal societies as a result of European-induced culture conflict. Man (n.s.) 23:654-670.

Brock, L. 1991. Peace through parks: the environment on the peace research agenda. Journal of Peace Research 28(4):407-423.

Brown, M.W., M. Kaku, J.M. Fallows, and E. Fischer. 1991. War and the environment. Audubon 93(5):88-99.

Chagnon, N.A. 1988. Life histories, blood revenge, and warfare in a tribal population. Science 239:985-992.

CIMI (Conselho Indigenista Mission). 1987. Doctrine of national security threatens Brazil's Indians. Cultural Survival Quarterly 11(2):63-65.

Conklin, B.A. and L.R. Graham. 1995. The shifting middle ground: Amazonian Indians and ecopolitics. American Anthropologist 97(4):695-710.

Crosby, A. 1986. Ecological Imperialism. Cambridge University Press, New York.

Dillon, T.C. and E.D. Wikramanayake. 1997. Parks, peace and progress: a forum for transboundary conservation in Indo-China. PARKS 7(3):36-51.

Edgerton, R.B. 1992. Sick Societies: Challenging the Myth of Primitive Harmony. The Free Press, New York.

Ember, C.R. and M. Ember. 1992. Resource unpredictability, mistrust, and war. Journal of Conflict Resolution 36(2):242-262.

Faiola, A. 1998. Peru, Ecuador sign pact ending border dispute. The Washington Post, 27 October.

Fairhead, J. and M. Leach. 1995. False forest history, complicit social analysis: rethinking some West African environmental narratives. World Development 23(6): 1023-1035.

Ferguson, R.B. 1989a. Ecological consequences of Amazonian warfare. Ethnology 28:249-264.

Ferguson, R.B. 1989b. Game wars? Ecology and conflict in Amazonia. Journal of Anthropological Research 45:179-206.

Flannery, T. 1994. The Future Eaters: An Ecological History of the Australasian Lands and People. George Braziller, New York.

Hanks, J. 1998. Protected areas during and after conflict: the objectives and activities of the Peace Parks Foundation. PARKS 7(3):11-24.

Harbinson, R. 1992. Burma's forests fall victim to war. The Ecologist 22(2):72-73.

Harris, M. 1974. Cows, Pigs, Wars and Witches: The Riddles of Culture. Random House, New York.
Hart, T.B. and J.A. Hart. 1997. Zaire: new models for an emerging state. Conservation Biology 11(2):308-309.
Heider, K. 1970. The Dugum Dani: A Papuan Culture in the Highlands of West New Guinea. Aldine, Chicago.
Hill, J.D. 1994. Alienated targets: military discourse and the disempowerment of indigenous Amazonian peoples in Venezuela. Identities 1(1):7-34.
Homer-Dixon, T.F. 1994. Environmental scarcities and violent conflict: evidence from cases. International Security 19(1):5-40.
ICPF (International Commission on Peace and Food). 1994. Uncommon Opportunities: An Agenda for Peace and Equitable Development. Zed Books, London.
IUCN (International Union for Conservation of Nature and Natural Resources) (ed.). 1997. Parks for Peace Conference Proceedings. IUCN, Gland, Switzerland.
Jobogo Mirindi, J.P. 1990. Violation des limites au parc national des Virunga. Southern Africa Wildlife College, Hoedspruit, South Africa.
Kane, H. 1995. The hour of departure: Forces that create refugees and migrants. WorldWatch Paper 125:1-56.
Keeley, L.H. 1996. War Before Civilization. Oxford University Press, New York.
Knauft, B.M. 1990. Melanesian warfare: a theoretical history. Oceania 60:250-311.
Lee, K.N. 1993. Compass and Gyroscope, Integrating science and politics for the environment. Island Press, Washington DC.
Lewis, D. 1990. Brazil's army loses temper. BBC Wildlife, July:483.
Martin, P.S. and R.G. Klein (eds.). 1984. Quaternary Extinctions: A Prehistoric Revolution. University of Arizona Press, Tucson.
McClure, H.E. 1974. Migration and Survival of the Birds of Asia. US Army Component, Seato Medical Research Laboratory, Bangkok.
McNeely, J.A. and P.S. Wachtel. 1988. Soul of the Tiger: Searching for Nature's Answers in Southeast Asia. Oxford University Press, Singapore.
McNeely, J.A., K.R. Miller, W.V. Reid, R.A. Mittermeier, and T.B. Werner. 1990. Conserving the World's Biological Diversity. IUCN, Gland, Switzerland; WRI, CI, WWF–US, and the World Bank, Washington DC.
McNeely, J.A. 1994. Lessons from the past: forests and biodiversity. Biodiversity and Conservation 3:3-20.
Meggitt, M. 1977. Blood Is Their Argument: Warfare Among the Mae Enga Tribesmen of the New Guinea Highlands. Mayfield, Palo Alto.
Myers, N. 1979. Wildlife and the dogs of war. The Daily Telegraph (London), 8 December.
Nietschmann, B. 1990a. Battlefields of ashes and mud. Natural History 11:35-37.
Nietschmann, B. 1990b. Conservation by conflict in Nicaragua. Natural History 11:42-49.
Orians, G.H. and E.W. Pfeiffer. 1970. Ecological effects of the war in Vietnam. Science 168:544-554.
Pearce, F. 1994. Soldiers lay waste to Africa's oldest park. New Scientist, 3 December:4.

Ponting, C. 1992. A Green History of the World: The Environment and the Collapse of Great Civilizations. St. Martin's Press, New York.
Renner, M. 1996. Fighting for Survival: Environmental Decline, Social Conflict, and the New Age of Insecurity. W.W. Norton and Co., New York.
Shankman, P. 1991. Culture contact, cultural ecology, and Dani warfare. Man (n.s.) 26:299-321.
Strada, G. 1996. The horror of land mines. Scientific American 274(5):26-31.
Thacher, P. 1984. Peril and Opportunity: What it takes to make our choice. Pp. 12-14 in McNeely, J.A. and K.R. Miller (eds.) National Parks, Conservation, and Development: The Role of Protected Areas in Sustaining Society. Smithsonian Institution Press, Washington DC.
Thorsell, J. (ed.). 1990. Parks on the Borderline: Experience in Transfrontier Conservation. IUCN, Gland, Switzerland.
Vayda, A.P. 1974. Warfare in ecological perspective. Annual Review of Ecology and Systematics 5:183-193.
Westing, A.H. 1976. Ecological Consequences of the Second Indo-China War. Almqvist and Wiksell, Stockholm.
Westing, A.H. 1982. The environmental aftermath of warfare in Vietnam. Pp. 363-389 in SIPRI. World Armaments and Disarmament: SIPRI Year Book 1982. Taylor and Francis Ltd., London.
Westing, A.H. 1993. Transfrontier reserve for peace and nature on the Korean Peninsula. Pp. 235-242 in IUCN (ed.). Parks for Peace Conference Proceedings. IUCN, Gland, Switzerland.
Westing, A.H. 1998. Establishment and management of transfrontier reserves for conflict prevention and confidence building. Environmental Conservation. 25(2): 91-94.
Winter, P. 1997. Wildlife and war. Swara July/August:6-7.
Wolkomir, R. and J. Wolkomir. 1992. Caught in the cross-fire. International Wildlife 22(1):5-11.
Zbicz, D.C. and M. Greene. 1997. Status of the world's transfrontier protected areas. PARKS 7(3):5-10.

Contras and Comandantes: Armed Movements and Forest Conservation in Nicaragua's Bosawas Biosphere Reserve

David Kaimowitz
Angelica Fauné

SUMMARY. In the 1980s, Nicaragua's Sandinista government faced armed mestizo and indigenous insurgencies in much of the nation's central and eastern regions. After the Sandinistas lost the 1990 elections, the in-coming government signed peace agreements with the insurgents and facilitated their return to civilian life. With the war over, the Nicaraguan army greatly reduced its troop strength, leaving tens of thousands of people unemployed. Within a few years, however, many former insurgents and soldiers took up arms again for multiple and complex reasons. This paper examines how three groups that rearmed influenced forest conservation in the buffer zone of Nicaragua's Bosawas Biosphere Reserve between 1991 and 1999. The three groups were the mestizo Northern Front 3-80 (FN 3-80) and the Andrés Castro United Forces (FUAC), made up of former 'Nicaraguan Resistance' and Sandinista soldiers respectively, and the Miskito YATAMA movement. The presence of these armed groups impeded the government from taking coercive action to remove farmers from the reserve's nucleus. It also limited the advance of cattle ranching. At times, the groups favored logging, at times they did not. The

---

David Kaimowitz is Director General, Center for International Forestry Research, P.O. Box 6596 JKPWB, Jakarta 10065, Indonesia.

Angelica Fauné is an independent consultant, Apartado LM-184, Managua, Nicaragua.

[Haworth co-indexing entry note]: "Contras and Comandantes: Armed Movements and Forest Conservation in Nicaragua's Bosawas Biosphere Reserve." Kaimowitz, David, and Angelica Fauné. Co-published simultaneously in *Journal of Sustainable Forestry* (Food Product Press, an imprint of The Haworth Press, Inc.) Vol. 16, No. 3/4, 2003, pp. 21-47; and: *War and Tropical Forests: Conservation in Areas of Armed Conflict* (ed: Steven V. Price) Food Products Press, an imprint of The Haworth Press, Inc., 2003, pp. 21-47. Single or multiple copies of this article are available for a fee from The Haworth Document Delivery Service [1-800-HAWORTH, 9:00 a.m. - 5:00 p.m. (EST). E-mail address: getinfo@haworthpressinc.com].

armed conflicts have tended to keep out prudent foreign investors and encourage the presence of smaller companies willing to take greater risks.

KEYWORDS. Bosawas, forest conservation, Sandinistas, cattle ranching, conflict, guerrillas, logging, mining, Miskitos, Nicaragua, protected areas, war

INTRODUCTION

This paper considers how the presence of autonomous armed groups affected the management and use of forest resources in six municipalities that straddle the border between Nicaragua's predominantly mestizo interior and the indigenous territories of the Atlantic Coast. We chose these six municipalities–Cúa-Bocay and Wiwilí in the department of Jinotega and Bonanza, Siuna, Waslala, and Waspam in the North Atlantic Autonomous Region (RAAN)–because they include a large part of Central America's largest remaining rainforest. They also include the Bosawas Biosphere Reserve, the country's largest protected area (Figure 1).

Violence and armed conflict have become a way of life in northeast Nicaragua. Given the central government's limited effective presence in the area, many people there have decided to earn their money through brute force. Some defend their use of violence with sophisticated political rhetoric. Others do not bother to construct an argument, but just want money. Most cases lie somewhere in between. Former anti-Sandinista Nicaraguan Resistance fighters, Miskito combatants, and Sandinista soldiers have risen up in arms because they feel neglected and abandoned by the government. They resent that others have greater access to the region's natural riches, government funds, and foreign money than they do. They also want other armed groups to stop harassing them. At times, they ally themselves with outside forces that have their own political agendas. Ultimately though, armed force is used to get support from the government, to establish control over the region's land, timber, and gold, and to take money, cattle, or food from their neighbors.

During much of the 1990s, three armed groups controlled significant

FIGURE 1. Map of Nicaragua.

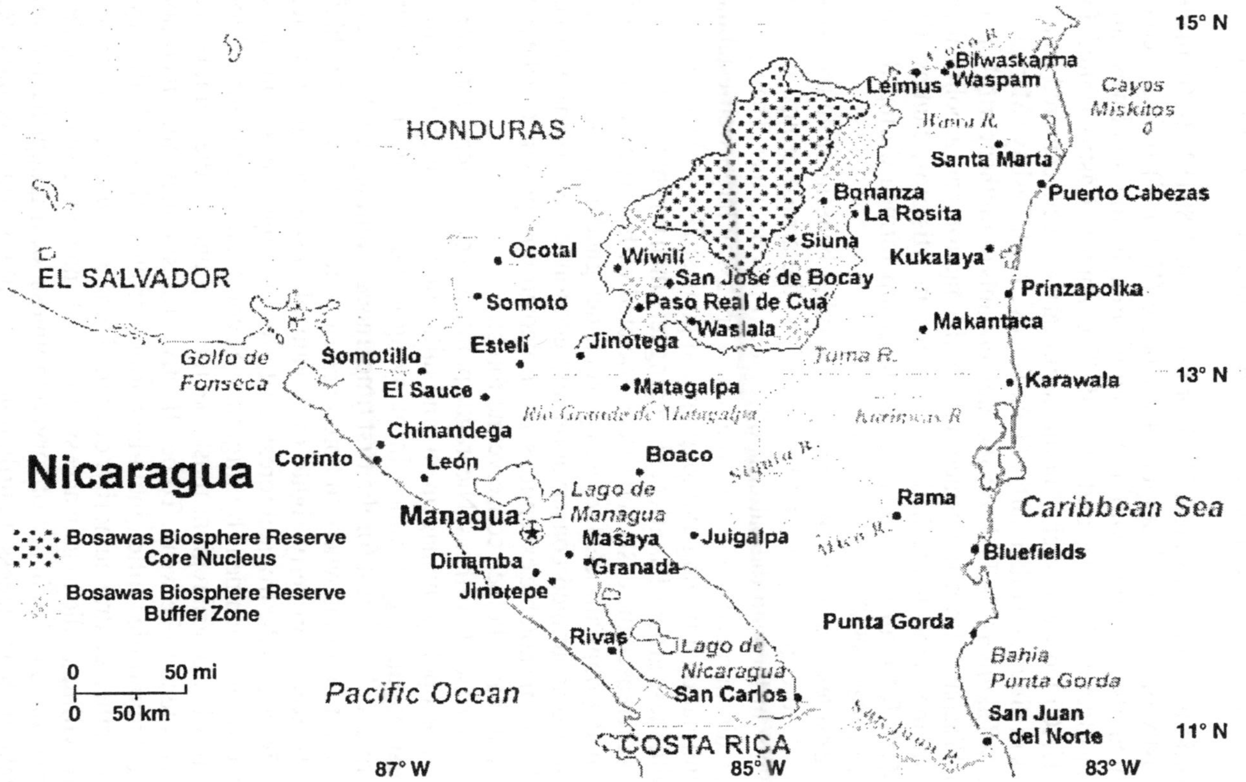

portions of the study area. The Northern Front 3-80 (FN 3-80), made up mostly of former Nicaraguan Resistance ('Contra') fighters, concentrated its activities within the study area in the southwest around Cúa-Bocay, Waslala, and Wiwilí. The Andrés Castro United Front (FUAC) led by former Sandinista military officers centered its activities in Siuna and other parts of Cúa-Bocay. Yapti Tasbaya Maraska nani Asla Takanka (YATAMA), the 'descendants of mother earth,' operated in Waspam. Both the armed YATAMA faction and a political party with the same name had their origins in Miskitos, Sumos, Ramas (MISURA), an indigenous guerrilla group formed during the 1980s (Gonzalez 1997).

Besides extorting land, money, and other benefits from the government, foreign donors, and local farmers, the armed groups have indirectly influenced the patterns of regional resource use. Their presence has discouraged cattle ranching and kept away large well-established multinational mining and logging companies. The violence associated with these groups has affected logging less than ranching, but even so, it has increased the costs and risks associated with it. The existence of the armed groups attracted foreign projects designed to 'pacify' the region, but it discouraged conservation and economic development projects. With some notable exceptions, the armed groups' presence in the region has probably helped conserve the forest.

The case of northeast Nicaragua points to more general issues related to forest policy, management, and utilization in developing countries. First, violence is common in many agricultural frontier areas where farmers clear forest for the first time. Indeed, military conflicts are endemic to regions where the presence of central government is historically weak, property rights over valuable resources are not clearly defined, and the tropical forest territories of indigenous groups are subject to encroachment by migrants. Second, any discussion of forest policy in tropical regions must take into account that the central governments tend to be weak. To be effective, forest policy must incorporate local authorities (often including armed groups), find realistic ways to strengthen central authority, or both. Third, in situations where people are willing to die to gain access to valuable natural resources, protected areas have little chance of surviving unless local groups have a strong stake in their success. Fourth, unless central governments and international agencies address property rights and governance issues in forested regions, these problems can grow until they explode, causing major political and economic crises that reach far beyond the forest.

The paper first provides basic background on the military conflicts in Nicaragua during the 1980s and their effects on the study area. Then it presents basic background on the study area itself and on the Bosawas Biosphere Reserve. After that, it discusses the FN 3-80, the FUAC, and the YATAMA and their impact on the forest, before concluding.

WAR AND PEACE IN THE 1980S IN NORTHEAST NICARAGUA

The Sandinista Front for National Liberation (FSLN) seized power in July 1979. Soon afterwards, the Sandinistas faced anti-government insurgencies in the country's interior and Atlantic Coast. The Nicaraguan Democratic Force (FDN), more commonly known as the Contras, maintained a strong presence in much of that area. It brought together former members of Nicaragua's National Guard and disgruntled mestizo farmers. MISURA, the predecessor of YATAMA, was composed mostly of Miskito Indians and operated on the Atlantic Coast.

The Mestizo FDN Insurgency

Most farmers in the recently colonized agricultural frontier areas of Jinotega and nearby Matagalpa initially hailed the Sandinistas' victory. However, when the Sandinistas imposed controls on food markets and expropriated landholders with strong local ties, some farmers turned against them and took up arms (Bendaña 1991). The Reagan administration in the United States directly supported the anti-government rebels. The Sandinistas responded to the insurgency with repression, which further fueled the revolts. Through this process, the FDN was transformed from a small group of former National Guard troops and mercenaries into a veritable campesino army. To defend themselves from the Contras' attacks, many farmers who remained loyal to the Sandinistas found themselves obliged to join the Sandinista army and militias.

Between 1983 and 1989, the FDN dominated the rural areas in most of the region covered by this study. The government controlled the towns and the parts of Cúa and Siuna where most Sandinista agricultural production cooperatives were located, but little else. MISURA, which will be discussed below, held most of Waspam. Fighting was fierce and many people died. Thousands of farmers fled their villages, and thick brush and young secondary forests developed in the abandoned fields and pastures. Farming, logging, mining, and hunting either ceased or greatly diminished because of the war, and the environment consequently benefited (Nietschmann 1990).

Militarily, the FDN was no match for the Sandinista army. Nonetheless, the 'low intensity' war they conducted eventually took a harsh economic and human toll. By 1987, the effects of the war, combined with ill-conceived economic policies and an economic embargo imposed by the United States, had pushed Nicaragua's economy into a severe recession. As the years passed, the population became war worn and the Sandinista leadership realized they had to seek a negotiated solution. This coincided with shifts in the international arena that opened fresh opportunities for compromise. Consequently, the Sandinistas began negotiating with both their Central American neighbors and the Contras. This culminated in the 1990 elections, where opposition candidate Violeta Barrios de Chamorro defeated Sandinista leader Daniel Ortega and assumed the presidency.

The Miskito MISURA Insurgency and Its Historical Origins

The Miskito insurgency shared similarities with the FDN's mestizo rebellion, but also had important differences. As with the mestizos, the Miskitos' initial sympathy for the Sandinista government soon soured. U.S. support for armed resistance, combined with a spiral of protest and repression, led to widespread conflict. However, unlike the mestizo colonization areas, which were linked by road to Nicaragua's Pacific Coast, the Miskito regions had historically been quite isolated from the rest of Nicaragua and their insurgency had strong ethnic underpinnings.

Beginning in the seventeenth century, Britain dominated the Atlantic Coast for some two hundred years. For the most part, it ruled indirectly and permitted the Miskito Indians and other local inhabitants to manage their own affairs. Around 1680, the British crowned a Miskito king and recognized the Mosquito Kingdom as the local government. Then, between 1787 and 1844, the British formally recognized Spanish authority in the region. However, strong Miskito resistance kept the Spanish from governing and the British reasserted control until 1860, when they officially recognized Nicaraguan sovereignty over the region. Even then, Nicaragua agreed to establish its own 'Miskito Reserve' on the Atlantic Coast, with its own constitution and an English legal system (Hale 1994).

Nicaragua did not attempt to govern the Atlantic Coast until 1894, when President José Santos Zelaya sent troops there and abolished the

Miskito Reserve. Zelaya forced the Miskito king into exile, named a mestizo governor, and declared Spanish the official language (Hale 1994). However, before he could effectively impose his rule in the region, the United States government forced him out of office in 1909. During most of the next twenty-five years, the country was racked by civil war and occupied by U.S. Marines, and the Nicaraguan government mostly ignored the Atlantic Coast.

Under the Somoza dynasty that lasted between 1934 and 1979, the central government had minimal influence on the Atlantic Coast. The Nicaraguan National Guard maintained only a nominal presence. In the Miskito villages along the Coco River, the Moravian Church assumed traditional government functions such as the provision of education and health care. North American mining companies provided electricity, water, health care, and other services in Bonanza and Siuna. Elsewhere people relied on their own resources.

When the Sandinistas took power in 1979, they applied the same policies and practices in the Atlantic and Pacific Coasts. Consequently, education, health care, and access to credit greatly improved. However, they also imposed measures and authorities without taking into account the cultural differences between the two regions. Overnight, the situation on the Atlantic Coast went from (not so benign) neglect to (even less benign) massive government intervention. The government sent thousands of teachers, doctors, soldiers, and administrators to the Atlantic Coast and established government offices and mass organizations modeled after those on the Pacific Coast. The new arrivals rarely spoke the local languages (English, Miskito, Mayangna, Rama, and Garifona), came from heavily Catholic backgrounds (whereas the Moravian Church was the most important on the Coast), and frequently expressed racist sentiments about the local population. Their strident revolutionary rhetoric did not match the Coast's historical experience. Matters were not helped when the Sandinistas nationalized the mines, fishing, and forest industries, but lacked sufficient funds and expertise to keep them running.

This situation coincided with and encouraged the gradual maturation of a militant ethnic consciousness among the Miskitos and a growing demand for regional autonomy. MISURASATA (Miskitos, Sumos, Ramas, and Sandinistas Working Together), an indigenous organization formed several months after the Sandinistas came to power, gave political expression to these demands. Although the Sandinistas supported the group's creation, relations between the two soon deteriorated. The Sandinistas began to harass and jail certain MISURASATA leaders, accusing them of secretly promoting an independent Miskito nation (Hale 1994).

In response, several Miskito leaders fled to Honduras and set up the MISURA guerrilla group. The Reagan administration provided MISURA with arms, money, and training in order to weaken the Sandinistas militarily and damage their international public image. This led to a familiar cycle of government repression, Miskito support for the guerrillas, and more repression. Within a few years, practically the entire Miskito and Mayangna population had fled to Honduras or been relocated, forcibly or voluntarily, to resettlement camps by the Nicaraguan government (Centro de Apoyo a Programas y Proyectos [CAPRI] 1998). Of the six municipalities discussed in this paper, Bonanza and Waspam were the most affected by the indigenous uprising.

Then came an extended period of negotiations. Realizing that their position was militarily and politically untenable, the Sandinistas backtracked and offered the Atlantic Coast's leaders major concessions, including regional autonomy. Out of this process emerged the 1987 Atlantic Coast Autonomy Law 28. It established two separate autonomous regions, each with a multi-ethnic government, and gave those governments substantial authority over their affairs (Acosta 1996). Four of the six municipalities in the study region–Bonanza, Siuna, Waslala, and Waspam–became part of the North Atlantic Autonomous Region (RAAN). These and other reconciliation measures contributed to a more favorable atmosphere for negotiations between the Nicaraguan government and the insurgent Miskito organizations. By then, MISURA had evolved into several factions, the largest of which was YATAMA.

Peace and Resettlement

Within months after taking office in April 1990, the government of President Violeta Barrios de Chamorro signed peace agreements with the mestizo and indigenous guerrilla groups still in arms. Consequently, 22,000 former FDN insurgents and sympathizers (now referred to as the Nicaraguan Resistance) and an unknown number of YATAMA fighters laid down their arms. Under the auspices of the International Commission for Support and Verification of the Organization of American States (CIAV-OAS) and the United Nations Observer Group in Central America (ONUCA), the ex-combatants and their families were resettled in a number of 'development poles' and security zones (Cuadra and Saldomando 1998). Most of these poles and zones were located in large expanses of unclaimed forest near the agricultural frontier areas where the Nicaraguan Resistance (RN) forces had operated. Ayapal in Cúa-Bocay and El Naranjo in Waslala, both near what would later become the Bosawas Reserve, were two cases in point (Stocks 1998). The government did not consider how resettling these soldiers near forested areas might affect the environment.

With the war over, Nicaragua reduced the size of its army, leaving tens of thousands of soldiers out of work. To compensate the former soldiers for their services and to avoid further social unrest, the government resettled many of them in frontier areas. Discharged officers received particularly large blocks of land. Other soldiers returned to where they came from or migrated to the agricultural frontier. Many eventually relocated to Siuna, particularly around El Hormiguero, a rural community adjacent to the Bosawas Reserve (Stocks 1998).

THE 'BOSAWAS' REGION AND THE BIOSPHERE RESERVE

The Regional Landscape

The six municipalities discussed in this paper extend from Nicaragua's mountainous interior to the Atlantic Coast. Together, they cover 22,946 km^2, an area roughly the size of El Salvador. Moving eastward, the elevation descends from about 600 m above sea level (masl) down to sea level. The two highest peaks, the Kilambé Mountain in Wiwilí and Peñas Blancas in Cúa-Bocay, tower over the region's western fringes. Each reaches around 1,750 masl, and the surrounding areas offer excellent conditions for coffee cultivation. The Isabelia Mountains traverse the region diagonally from Peñas Blancas to the northeast. Most of these mountains have peaks below than 900 masl. A large portion of Cúa-Bocay and northern Wiwilí also lies above 500 masl. The topography of these areas ranges from rolling hills to steep slopes. Outside these areas, the landscape becomes mostly flat tropical lowlands.

A dense network of rivers, streams, and creeks flows from the mountains to the Atlantic Ocean. Historically, the Amaka, Bocay, Coco, Lakus, Wina, and Waspuk Rivers formed the central axes of traditional indigenous settlements in the area. The name Bosawas itself comes from the first letters of the BOcay River, the SAslaya Mountain, and the WASpuk River.

As one travels eastward and into higher elevations precipitation increases steadily. Yearly rainfall averages between 1,600 and 2,000 mm in the western areas, but rises to over 3,000 mm in some of the eastern areas and higher elevations. Maize and beans attain greater yields in the drier areas, while rice and cocoa are better adapted to the areas with more rainfall. The higher areas to the west have fertile volcanic soils, and the alluvial soils of the lowland riverbanks are also quite rich. Indigenous agriculture has flourished for centuries along the rivers, but most other areas have poor, acidic, and heavily-weathered tropical soils.

Together with the adjoining territory of southeastern Honduras, this region has the largest remaining tropical rainforest in Central America. Outside eastern Waspam, where Caribbean pine (*Pinus caribeae*) forests constitute the natural vegetation, broadleaf forests originally covered almost the entire territory. These forests still house a large percentage of the country's 2,500 tree species, including highly coveted species such as mahogany (*Swietenia macrophylla*), Spanish cedar (*Cedrela odorata*), and "andiroba" (*Carapa guianensis*). They also constitute the habitat for a diverse wildlife, including: jaguars (*Felis onca*), monkeys, deer (*Mazama americana*), tapirs (*Tapirus bairdii*), caiman (*Caiman crocodilus*), parrots (including *Ara ambigua*), toucans (*Ramphastos* spp.), and various raptors.

The Region's Population and Production Systems

Small-scale mestizo farmers and ranchers dominate the western municipalities of Cúa-Bocay, Siuna, Waslala, and Wiwilí, where scrubby pasture lines most of the roads. In contrast, Waspam's population is almost entirely Miskito Indian, and the town has a distinctively Atlantic Coast indigenous culture. On the other hand, rural Bonanza and northern Cúa-Bocay and Wiwilí are similar. The town of Bonanza has all the trappings of an old mining settlement. Its small wooden houses are packed together along the main streets and daily life revolves around the Greenstone Mining Company.

Until the 1950s, no road reached the region, and it could be accessed only by boat or small plane. The government then built a road connecting Siuna to the Atlantic Ocean to encourage mining. About a decade later, the American Neptune Mining Company extended that road to Bonanza. Although the towns of Cúa and Wiwilí have been accessible by four-wheel drive vehicles from the Pacific Coast for some time, Bonanza and Siuna could not be reached by road from the Pacific until the 1970s. Even today, most places west of the town of Waspam are reached only by canoe or on foot.

Forty years ago, the region had less than 70,000 people. The population dropped to only a fraction of that during the war of the 1980s. However, by 1995 it had swelled to around a quarter of a million, principally thanks to the return of the displaced people and the arrival of many new migrants from other regions. In Cúa-Bocay, Siuna, and Wiwilí, the municipalities with the greatest in-migration, annual growth rates averaged 6% or more between 1971 and 1995. Only Waspam did not receive many migrants from the west. Indigenous people still represent 86% of the population, and less than 1% of the municipal population was born outside the Atlantic Coast (Baumeister 1997).

The region's main productive activities are agriculture, logging, and mining. In the mestizo areas, farmers raise cattle and grow corn, beans, rice, and coffee. Farms are typically around 50-100 ha, although larger and smaller farms exist. In the less conflictive areas near Peñas Blancas and Wiwilí, wealthier farmers have invested in coffee growing, but they have been reluctant to invest in ranching due to widespread violence, insecure land tenure, and the limited availability of credit. Mayangna and Miskito households have much more diversified livelihoods. Some own cattle and most grow a greater variety of crops, including more plantains, tubers, and rice. They also hunt and fish more than the mestizo communities. But, both mestizo farmers and the indigenous people engage in timber harvesting and pan for gold (Stocks 1998).

Before 1979, major multinational mining companies operated in Bonanza and Siuna. However, due to the violence and political instability that has characterized the region in recent years they have been reluctant to return. At present, Greenstone, a small Canadian mining company, is the only multinational mining company in operation. A number of smaller foreign-owned mining companies have mining concessions, but none is actively mining its claim.

Apart from the activities of local small-scale timber harvesters, outside loggers regularly enter the area, mostly looking for mahogany and cedar. In the Miskito areas along the Coco River, Maderas y Derivadas de Nicaragua S.A. (MADENSA), a Nicaraguan timber company established with investment from the Dominican Republic (Anaya and Crider 1998), practically monopolizes the timber trade. Sometimes the company harvests its own timber, but usually it purchases timber from local farmers. Other small multinational timber companies have tried to get a foothold in the area, but they have all failed because of local and national opposition. Wealthy Nicaraguan timber merchants dominate the trade in most of the region. They generally buy boards from small farmers who live near the Bosawas Reserve and extract timber from inside it.

On occasion, the merchants also hire their own logging crews or purchase wood from the reserve's inhabitants.

As mentioned previously, during the war of the 1980s, farming, mining, and logging came to a virtual halt. That allowed some timber resources and degraded soils to recuperate. The combination of government infrastructure investments and foreign-funded resettlement and conservation projects greatly favored the rapid conversion of forest to pasture and cropland, as well as the exploitation of the region's timber and gold. As will be shown below, however, the presence of the armed movements in the region significantly affected these processes.

The Bosawas Biosphere Reserve

In October 1991, approximately one and a half years after the official end of the war between the Nicaraguan government and the various anti-Sandinista insurgent groups, President Violeta Barrios de Chamorro signed Decree 44-91, creating the 'Bosawas National Resource Reserve.' That decree established a protected area of 7,400 km^2 within the study area, making it the largest protected area in Central America. Jaime Incer Barquero, the Minister of the Environment and the intellectual author of Decree 44-91, understood that with the end of the war the forests of the Bosawas region would come under great pressures from mining and timber companies and agricultural colonists. The creation of a protected area was intended to stave off these pressures. Nevertheless, at least in the beginning, the Bosawas National Resource Reserve was a paper park.

Incer Barquero was able to discourage certain mining and timber concessions from being made in the protected area, but he had practically no resources to invest in the area. He also lacked sufficient political power to influence the resettlement of former Sandinista and RN troops in the region. To obtain the necessary resources, he negotiated agreements with the United States Agency for International Development (USAID) and the German Technical Cooperation Agency (GTZ) to fund conservation projects in the region. GTZ provided US$3.8 million, and USAID–in partnership with The Nature Conservancy (TNC)–committed an additional US$2.5 million. However, practically none of that money arrived until 1994 (Ramírez et al. 1994). In the meantime, as will be discussed below, armed insurgents moved into much of the southern and western portions of the reserve.

The original Natural Resource Reserve was a heavily forested area, with a low population density, and no major town within its boundaries.

Eighty percent of the area belonged to the municipalities of Cúa-Bocay, Waspam, and Wiwilí, while the municipalities of Bonanza and Siuna held the remaining 20% (pers. comm., A. Fauné 1999). As of the mid-1990s, about 25,000 people (more or less equally divided between indigenous people and mestizos) lived in the area. Over two-thirds of the mestizos moved into the area after 1990, settling mostly in the south along the Bocay, Iyas, and Wina Rivers (Stocks 1998).

During the first years after the official creation of the reserve, most its inhabitants were unaware or only vaguely aware that they lived in a protected area. No one consulted them before creating the reserve (Howard 1997). In 1996, the Ministry of Environment and Natural Resources (MARENA) formally divided the reserve into a core nucleus of 3,300 km^2 where no human activities were allowed and a buffer zone where the local inhabitants could engage in their traditional activities, with certain restrictions (Secretaria Técnica de Bosawas [SETAB] 1996).

In 1998, the United Nations Educational, Scientific and Cultural Organization (UNESCO), in coordination with MARENA and the USAID and GTZ projects, declared the six municipalities and 22,946 km^2 addressed in this paper a 'World Biosphere Reserve.' In keeping with the new classification, the Nicaraguan government changed the name of the Bosawas Natural Resource Reserve to the Bosawas Biosphere Reserve (CAPRI 1998). No additional restrictions on land use in the area outside the initial Natural Resource Reserve accompanied the creation of the Biosphere Reserve. Hence, in the remainder of the paper, the term 'the Bosawas Biosphere Reserve' refers to the 7,400 km^2 where the government has officially placed restrictions on land use. Before discussing the activities of the armed movements in the region, it is worth noting that although only 20% of 7,400 km^2 Bosawas Biosphere Reserve belongs to the municipalities of Bonanza and Siuna, both the GTZ and USAID projects concentrated their efforts there. One major reason for this has been ongoing violence in the municipalities of Cúa-Bocay, and to a lesser extent, Waspam.

THE THREE ARMED MOVEMENTS AND THE FOREST

The following sections discuss three armed movements in the Bosawas region and their impact on forests. They provide information on who led and composed the movements, the major events in their history, what motivated their actions, and how they influenced the production systems in the areas where they operated. At different times, these movements effectively governed portions of the territory and therefore determined natural resource and environmental policy and regulation in those areas. Although the discussion focuses on the groups' activities within the study area, two of the armed movements analyzed–the Northern Front 3-80 and the YATAMA–also operated in nearby municipalities outside the study area.

The Northern Front 3-80

After the Nicaraguan Resistance (RN) laid down its arms in 1990, the Nicaraguan government earmarked Siuna and Waslala as sites for 'development poles.' The government chose these areas because they still contained available land. In addition, many RN fighters came from or had operated in the area (Cuadra and Saldomando 1998). The CIAV-OAS was expected to oversee and assist the resettlement process (United States Department of State 1998).

Nevertheless, the government and foreign agencies proved incapable of meeting the former Contras' overwhelming demands for land, credit, and social services. Continuing incidents of violence between former Contras and ex-Sandinista soldiers gave many groups a reason to rearm. When it became obvious that the Chamorro administration was unwilling to confront the FSLN and its sympathizers in the Nicaraguan army, right-wing groups fostered the creation of armed bands to pressure it into doing so. These groups wanted President Chamorro to return properties confiscated by the Sandinistas to their original owners and eliminate all Sandinista presence in the armed forces and security apparatus.

Various armed organizations comprising former RN fighters emerged in this context. One of those groups was the FN 3-80, which took its name from the pseudonym used by Enrique Bermúdez, a Contra commander assassinated in 1990. Its leader José Angel Talavera, better known as the 'Jackal,' initially demanded that President Chamorro fire Minister of Defense Humberto Ortega and her son-in-law, Antonio Lacayo, the minister of the presidency. He also demanded measures to ensure the safety of former Contra fighters. Right-wing Nicaraguans and Cubans in Miami, conservative political forces within Nicaragua itself, and U.S. politicians such as Senator Jesse Helms in Washington, D.C. gave funds and political support to the FN 3-80 (Nicaragua Network 1993).

Between 1992 and 1997, the FN 3-80 exercised significant influence over large areas of Cúa-Bocay, Waslala, Wiwilí, and other nearby municipalities. On various occasions during that period, the FN 3-80 negotiated with the Nicaraguan government about disarming. As part of that negotiation process, the Nicaraguan army pulled out of certain areas, effectively leaving them in the hands of the FN 3-80. Therefore, the FN 3-80 was able to determine who had access to the natural resources within their territory and under what conditions. For example, when the FN 3-80 decided that it opposed efforts by the Nicarguan Institute of Agrarian Reform (INRA) to title land in the area, INRA was forced to suspend its program. Similarly, the CIAV-OAS and a European Union (EU) rural development project located in Cúa-Bocay regularly consulted with the FN 3-80's commanders before deciding who received their services and where their investments were made. Talavera's troops also maintained 'order' in the region, killing off thieves, cattle rustlers, and rapists, as well as assassinating several hundred Sandinista sympathizers.

In early 1993, negotiations between the Nicaraguan government and the FN 3-80 began in earnest. Then, in August of that year, the FN 3-80 kidnapped a government delegation that had traveled to the area with the purpose of convincing Talavera to disarm. In response, a pro-Sandinista force took hostages of its own, including Nicaragua's vice president, Virgilio Godoy. This led to a five-day standoff that ended when each side released its hostages. After that, the Nicaraguan army refrained from attacking the FN 3-80 for several months. When the negotiations broke down again, intense fighting followed. In early 1994, just when the Nicaraguan army was on the verge of defeating the FN 3-80, a cease-fire was proposed by Miguel Obando, Nicaragua's Cardinal, and Pablo Antonio Cuadra, the director of the conservative newspaper *La Prensa*. This forced the Nicaraguan army to step back. Soon after, the FN 3-80 and the Nicaraguan army declared a cease-fire. Following several weeks of negotiations, the government signed a disarmament agreement with Talavera on February 25, 1994 (Nicaragua Network 1994).

As part of the agreement, the government offered to provide land, credit, technical aid, and medical assistance, and to pay for each automatic rifle handed over. In addition, the CIAV-OAS promised to increase its activities in the areas controlled by the FN 3-80. The Nicaraguan army also agreed to limit its troop strength in eight towns, including San José de Bocay and Wiwilí, and allowed FN 3-80 members to assume high-level positions in the police departments of Bocay, Cúa, and Wiwilí (Nicaragua Network 1994).

By April, several hundred FN 3-80 members had disarmed and moved into security zones in El Naranjo and Kubalí (in Waslala) and

Ayapal (in Cúa-Bocay). However, a number of FN 3-80 commanders, including Sergio Palacios ('El Charro') refused to disarm and went off on their own. Unlike Talavera, these commanders were less interested in politics and not as well connected with right-wing political groups. They simply felt there was not enough in the deal for them (Cuadra and Saldomando 1998).

The presence of the FN 3-80 during this entire period greatly facilitated the entrance of former RN soldiers and other mestizo farmers into the area along the Bocay River and the Wina River north of Ayapal and along the Iyas River north of El Naranjo. These areas formed the southern fringes of the Bosawas Biosphere Reserve. As Anthony Stocks, an American anthropologist working in the area put it, "In a practical sense as well as a kinship sense these guerillas are just another face of the land invasions of mestizos settlers into the Biosphere Reserve" (1995, 13). Although the government had prohibited families from moving to these areas, such prohibitions meant little since these areas remained beyond the control of the government and the Nicaraguan army. As noted previously, given the FN 3-80's presence in the area neither of the conservation projects in the Bosawas Biosphere Reserve was willing to work in the southern portions of the reserve.

The logging situation was more complex. Around 1995, the EU-financed rural development project in Cúa-Bocay improved the road leading to Ayapal, and the loggers followed. 'El Charro' allowed them to work in the area as long as they gave him money, boots, and other provisions. In Waslala, he went so far as to issue his own 'logging permits.' According to one person interviewed, the most intense logging took place in Waslala while the FN 3-80 effectively controlled the area between 1993 and 1995. Another person commented that 'El Charro's supporters were willing to protect the loggers in return for a few cartons of cigarettes. In the El Naranjo security zone, many people began working as chainsaw operators. Around Bocay, the FN 3-80 served as bodyguards for the loggers and threatened or attacked anyone who opposed their activities (Comisión Nacional de Bosawás 1995). For reasons that remain slightly unclear, 'El Charro' became increasingly concerned about the loggers' negative impact on the environment. Apparently, a prominent local environmentalist convinced him that logging would provoke droughts and dry up local water sources. Some of his local supporters also opposed the loggers and even went so far as to destroy some of their tractors and equipment.

Although the FN 3-80 favored cattle ranching, and most of its members aspired to become ranchers, their activities discouraged investment in ranching. FN 3-80 fighters frequently took and ate cattle, and their presence made ranchers wary to enter the area. This helps to explain why cattle ranching advanced more slowly in this area than in other agricultural frontier areas of Nicaragua during the 1990s.

Between mid-1994 and mid-1996, there were occasional skirmishes between the FN 3-80 fighters and the Nicaraguan army. FN 3-80 troops continued to kill, rob, and rape local residents and terrorize anyone they believed supported the Sandinistas (González Silva 1997). The Nicaraguan army then scored a major victory when it killed 'El Charro' and one of his top lieutenants in Waslala (United States State Department 1998).

'El Charro's death opened the door to a new set of negotiations. In May 1997, the two parties reached an agreement that included land, food, clothing, seeds, housing materials, and services for FN 3-80 members, as well as amnesty and security guarantees (Policy, Planning and Research Branch, Research Directorate, Immigration and Refugee Board of Canada [DIRB] 1997). At the request of FN 3-80 commanders the government added a clause whereby "both parties agree to combat the destruction of the forests and the government promises to take the necessary steps to avoid their depletion." Unofficially, the government allowed the FN 3-80 to name the auxiliary mayors of Ayapal and El Naranjo. By the time President Arnoldo Aleman flew into Ayapal by helicopter in July 1997 for the ceremony formalizing the completion of the disarmament, 1,197 members of the FN 3-80 had laid down their arms (Associated Press 1997). Since the FN 3-80 officially disarmed, a number of former members have joined criminal bands and taken to random kidnappings, assaults, and cattle theft. This has continued to discourage investment in cattle ranching.

The Andrés Castro United Front (FUAC)

After the defeat of the Sandinistas in the 1990 elections and the subsequent demobilization of large numbers of soldiers from the Nicaraguan army, many of those soldiers were just as unsatisfied as their Contra counterparts. Many felt their revolution was being betrayed, as the government returned confiscated properties to their previous owners and former sympathizers of the Somoza regime returned to the country. They saw the high command of the Nicaraguan army turning its back on former colleagues in order to protect its own interests and those of the new propertied classes. Moreover, they found themselves with little land, credit, or other forms of government support. Some responded by taking up arms.

Within the study area, the main group that fit this description was the Andrés Castro United Front (FUAC), named after a young Nicaraguan who fought against the American filibuster William Walker in the middle of the last century (González Silva 1997). Former military officers clandestinely formed the FUAC in Managua sometime around 1992. Its founders had substantial combat experience and knew how to build a network of logistical and political support. The FUAC's two main leaders were Edmundo García (a.k.a. 'Camilo Turcios') and Gustavo Navarro (a.k.a. 'Tito Fuentes').

Several years after the FUAC's formation, the group moved its base to Siuna. There, it concentrated its attention on the villages along the roads between Waslala and Siuna and between Siuna and Mulukukú in the southwest part of the municipality, and around Coperna in the southeast. It also moved into the area near the town of Cúa in Cúa-Bocay. These areas offered fertile ground for the FUAC. Many farmers there had received land from the Sandinistas during the agrarian reform in the 1980s, belonged to pro-Sandinista agricultural cooperatives, and had substantial combat experience fighting against the Nicaraguan Resistance. Under the Sandinistas, these farmers had been among those most favored by government policies. In contrast, the Chamorro and Aleman governments stopped giving them agricultural credit, cut funding for their health care and education, failed to maintain the local roads, and showed no interest in titling the cooperatives' lands. When the FUAC talked about the poor condition of the roads and the lack of transport, credit, electricity, and health care, they found many farmers willing to listen. The farmers were even more impressed when the FUAC began killing cattle rustlers and other criminals that had plagued them. Siuna's mostly forested landscape also facilitated the FUAC's operations, allowing it to hide from the Nicaraguan army.

From the beginning, the FUAC invoked a mix of broad political attacks against President Aleman, specific appeals for improvements in local conditions, and demands for material benefits for its own members (FUAC 1997a, 1997b). Sometimes it appeared to represent local farmers and the residents of Siuna. On other occasions, it focused solely on its own members, almost all of who were former soldiers.

The fifth point in the FUAC's list of demands called for, "the respect and conservation of national natural resources, which includes laws related to the exploitation and management of the same, taking into account the populations of the neighboring areas." To elaborate such laws and regulations, the FUAC proposed a "technical commission including environmental organizations and civil society producers and professional groups" (FUAC 1997a). The FUAC also called on the government to grant a forest concession large enough to create jobs for 500 local people to its proposed foundation. Interviewed by the authors in mid-1998, Turcios and Fuentes made it clear that while they objected to outsiders exploiting the region's resources, they had no problem with logging by local people. They stated that, "In the FUAC, we believe that the only ones who can save this region is its own population . . . They are the ones who can guarantee the sustainable management of their forest resources, of their riches that others are trying to snatch away from them" (1998, authors' translation).

In a bizarre incident in September 1997, a group calling itself the Ecological Armed Front (FEA) issued a communiqué saying it had "taken up arms to defend against the unscrupulous loggers who are principally responsible for the destruction of the environment." It proceeded to confiscate and burn 25 chainsaws in the central plaza of Puerto Viejo in Waslala "as a warning against people and companies that dedicate themselves to cutting down forests and destroying natural resources." The group also declared that it wanted to end government corruption and said it would not respect government logging licenses (Nicaragua Network 1997). The FEA was never linked to the FUAC, but given that the FUAC essentially controlled Puerto Viejo at the time, it is plausible that the FUAC at least acquiesced to FEA's actions.

At the same time, the FUAC apparently chose to ignore the destructive logging activities of a group made up predominantly of former Sandinista army officers that called itself the Agro-Forestry Cooperative. The Agro-Forestry Cooperative operated extensively near FUAC strongholds and probably could not have done so without tacit FUAC approval. The FUAC also strongly supported road improvement, with little regard for the potential impact on forests. When the mayor of Cúa-Bocay failed to uphold his promise to build a road, the FUAC kidnapped him. They also demanded that the government improve the road from Waslala to Siuna, part of which runs almost directly along the southeast edge of the Bosawas Biosphere Preserve.

In February 1997, the mayor of Siuna organized a peace commission, including the Nicaraguan army and the national police, to dialogue with the FUAC. The negotiations broke down and things reached a standoff. The mayor then convoked a civilian peace commission composed of representatives from local churches and non-governmental organizations. Despite the commission's efforts, several months of violence followed. In April 1997, the FUAC surrounded Rosita, a town close to Bonanza and Siuna, and threatened to capture it. Two days later, the Nicaraguan army attacked FUAC troops not far from where the peace negotiations were supposed to begin, and fighting continued for several months (DIRB 1997). Despite efforts by the mayor of Rosita to convince the government that foreign mining and timber companies would not invest in the area as long as the conflict continued, substantive negotiations did not begin until August. They finally reached fruition five months later. As part of the negotiation process that lasted from October to January, the FUAC moved its troops into four "peace enclaves," two of which bordered on the eastern edge of the Bosawas Biosphere Preserve. The FUAC had total control in the enclaves and was responsible for administering justice there.

As the negotiations proceeded, most general social and environmental demands fell by the wayside. Both parties increasingly concentrated on what the FUAC soldiers would receive. Concerning natural resources, the government agreed to set up a joint commission including MARENA, the Ministry of Defense, and the FUAC, but little happened after that. In the "final" peace accord signed by President Aleman and the FUAC in December 1997, the main government commitments were to provide FUAC members with land, health care, and scholarships, and to feed them for six months. The government also agreed to allow the FUAC to set up its own foundation and it promised to support the FUAC's efforts to secure international funding for housing, credit, infrastructure, and training (Government of Nicaragua 1997).

The FUAC officially disarmed on Christmas Day 1997, and 423 of its soldiers laid down their arms. However, within six months, the FUAC leadership was accusing the government of failing to meet its commitments. While the government had allowed the FUAC to set up their foundation and obtain foreign funding, it had largely failed to provide them with public services or land (Gomez Nadal 1997). Some former FUAC members took up arms again, calling themselves the Revolutionary Armed Forces (FAR). Others began talking about taking similar actions (Program for Arms Control, Disarmament, and Conversion [PACDC] 1998). In November 1999, these groups kidnapped a Canadian mining expert and several others. The Nicaraguan army accused Turcios of masterminding the kidnapping. Eventually, the kidnappers released the Canadian. Soon thereafter, unknown assailants killed both Turcios and Fuentes (Envío 2000). Violent bands with unclear motives continue to operate in the area.

Compared to the FN 3-80, the FUAC probably had a more limited impact on natural resource use in the region, at least initially. Their armed actions lasted only a few years. While they operated along the eastern and southeastern borders of the Bosawas Biosphere Reserve and may have facilitated illegal logging within the reserve, they apparently did not promote migration into the reserve as the FN 3-80 had. Nor did they regularly kidnap ranchers, like the FN 3-80. Indeed, the greatest impact on natural resource use in the area probably came from the more random criminal violence that erupted after the FUAC officially disbanded. That violence does seem to have discouraged investment in cattle ranching in the area and forced hundreds of farmers to flee their homes.

The YATAMA

Unlike the FN 3-80 and FUAC, the YATAMA had no central leader. It was a loose assembly of indigenous military commanders and their followers. After having what most considered negative experiences with the main indigenous *caudillos* during the 1980s, the local commanders were no longer willing to take orders from anyone. Each commander had his own view on different issues. Thus, it is more difficult to describe YATAMA's ideology. In fact, different sources refer to nine, thirteen, and twenty-five YATAMA demands to the government (Coca Palacios 1998).

The huge gap between the YATAMA's formal demands and the topics that eventually dominated their negotiations with the government further confuses the situation. The YATAMA agreed they were fighting for 'autonomy' and the 'demarcation of indigenous lands,' but what they meant remains unclear. They definitely did not feel that the 1987 regional autonomy law or the regional government provided the solution, and most of them did not particularly want land titles for their communities. They seemed to want the government to recognize one single large indigenous territory, to allow them to govern their own affairs directly, and to compensate them for the use of their natural resources. Above all, they were frustrated by their continuing poverty in a territory that was rich in timber, gold, fish, and other resources. Likewise, they resented that the central government discriminated against them despite their contribution to the defeat of the Sandinistas (Chamorro 1999). They felt that the Chamorro and Aleman governments paid less attention to them than to the former RN fighters, and on several occasions President Aleman had personally humiliated them. However, as with the FN 3-80 and the FUAC, the negotiations between the YATAMA and the government revolved around demands for housing, credit, food, and other types of direct government support.

Between 1992 and 1995, the YATAMA were involved in sporadic acts of violence. Each time, the government brought calm by promising small concessions, which it rarely delivered (Burke 1995; Cuadra Lira and Saldomando 1998). Natural resource issues played a minor role in this process. In 1992, the government promised to ask a U.S.-financed forestry project to hire demobilized YATAMA fighters as forest guards and train them in silviculture and forest management, as well as to turn over a small sawmill so Miskito families could construct their houses. In return, the demobilized fighters agreed to plant three trees for each one they cut (Hurtado et al. 1992). However, such agreements remained rather peripheral.

The conflict between the government and the YATAMA heated up again after February 1997, when the Miskito regional Council of Elders held its "IX General Assembly of Indigenous People and Ethnic Communities" in Puerto Cabezas. That assembly gave impetus to Miskito demands for regional autonomy and the demarcation of the indigenous territories. About a year later some 1,500 YATAMA fighters held their own assembly and decided to take up arms again (Flores 1998a). Within a short period, YATAMA troops controlled most of the towns along the Coco River and had attacked a government military post in Bismuna (Flores 1998b).

Negotiations followed almost immediately. The Nicaraguan army pulled its troops back from most of the Coco River and effectively allowed the YATAMA to control the region. This meant a large portion of the Bosawas Biosphere Reserve was in YATAMA hands. The army high command declared that the problem was political, not military, and said Aleman and his ministers should resolve it. While the negotiations continued, civilian government officials provided the YATAMA troops with food.

The negotiations did not go anywhere until July 1998, when the MADENSA sawmill in Puerto Cabezas burned down under mysterious circumstances (Leist 1998). Although the YATAMA never took credit for the fire, the pace of negotiations picked up soon afterwards. President Aleman and a few YATAMA commanders signed the first peace agreements in September 1998. The government promised to create offices in Puerto Cabezas and Waspam to provide the combatants with housing, credit, and land. It also agreed to help locate the bodies of Miskitos from the 1980s war, create a voluntary Miskito police force, and demarcate community lands. By February 1999, 1,500 of YATAMA's fighters had formally laid down their arms (Lopez 1999). Within a few months, however, as in the case of the FN 3-80 and the FUAC, the YATAMA were complaining that the government had failed to keep its promises, and talking about rearming again.

The MADENSA fire highlights the complex issue of the YATAMA's stance towards outside loggers. Interviews conducted by the authors with YATAMA commanders suggest that they did not object to outsider's logging in their territory, as long as they shared the benefits. Many Miskitos had previously worked for foreign logging companies and held positive feelings about their experience. However, they did reject the central government's claim to have the authority to grant logging concessions and licenses in what they considered their territory.

One "indigenous proclamation" that seven YATAMA commanders signed in May 1998 paints a more conservationist image than the previous paragraph suggests. It says, "Foreign companies and their concessions are freely destroying Mother Nature and its resources with the support of the government institutions and the regional governments. The forests disappear. The marine species are exterminated. The precious minerals are being depleted. The natural elements become scarce. The wild animals die and all the nature together with the Indians cries out in pain over the destruction." However, the authors are inclined to believe that this conservationist rhetoric does not fully reflect the YATAMA leaders' stance on forest resource exploitation, as most YATAMA commanders would accept outside logging companies in their territories as long as the communities benefited.

Whatever the YATAMA's feelings about logging, there is little doubt that as long as the Miskitos remain well armed and opposed to non-indigenous farmers moving into their territory, outside farmers are likely to think twice before doing so. Given that the Miskitos manage their resources reasonably well, this situation will probably benefit the forest. They maintain diverse cropping systems with long fallow periods and keep few cattle. While they may permit selective logging in their communities and engage in small-scale logging activities themselves, this would probably have limited impact on the forest, aside from depleting the commercial populations of mahogany and cedar.

CONCLUSIONS

The struggle to gain or maintain access to land and timber in the Bosawas region played a significant role in all three armed insurgencies examined in this paper. In this sense, conflict over natural resources directly contributed to violence in the region. Nevertheless, the groups proved more interested in receiving direct government support in the form of money and services.

The activities of the armed groups sometimes directly contributed to forest destruction. The presence of the FN 3-80 in the southern portion of the Bosawas Biosphere Reserve made it all but impossible for the central government to limit encroachment and logging there. In fact, their presence may have encouraged such activities. The FUAC tolerated and may even have facilitated illegal logging along the eastern edge of the reserve. All three groups used revenues from forestry and agricultural activities in or near the reserve to finance their activities. They were also unwilling to allow the government to impose restrictions on the use of natural resources where such restrictions would have negatively affected their local constituencies.

On the other hand, certain leaders of all three groups expressed concern at one time or another for the long-term sustainability of the forest resources. For reasons that are still not completely clear, all three armed groups raised the issue of natural resource conservation during their negotiations with the government. They were particularly interested in excluding logging companies that provided only limited benefits to local communities. In each case, they either directly attacked outside loggers or gave tacit support to others that did.

Independent of the armed groups' intentions regarding the exploitation or protection of natural resources, their mere presence discouraged investment by large cattle ranchers and well-established international mining and logging companies. The companies that moved into the area were mostly smaller operations willing to accept higher levels of risk. Likewise, small and medium ranchers were responsible for most of the livestock and pasture expansion in the area. Undoubtedly, the levels of investment by large companies and wealthy individual were also influenced by the general political and economic instability in the country, weak international markets, limited access to credit, and the region's poor infrastructure. Nevertheless, one should not underestimate the important role that violent conflict and the presence of armed groups played in deterring investment.

In the long term, the fact that the Miskitos remain heavily armed and firmly opposed to the arrival of non-indigenous farmers to the region may prove the most reliable deterrent to the advance of the agricultural frontier into their traditional territories. Although the YATAMA have formally disarmed, few believe that they have disarmed completely. It seems likely that endemic violence will continue to plague the Bosawas Region over the next few years, affecting both the mestizo and indigenous areas. Such violence may take the form of political or ethnic conflicts or just criminal behavior, but it will probably combine both. Little suggests that the effective government presence in the Bosawas Biosphere Reserve will increase, and the future of forest conservation in the region remains uncertain.

REFERENCES

Acosta, M.L. 1996. Los derechos de las comunidades y pueblos indígenas de la Costa Atlántica en la Constitución Política de Nicaragua y la implementación del estatuto de autonomía de las regiones autónomas de la Costa Atlántica de Nicaragua. Canadian International Development Agency, Managua.

Anaya, S.J. and S.T. Crider. 1998. The Mayanga Indigenous Community of Nicaragua: Moving from Conflict to a Convergence of Interests. pp. 202-222 in L.D. Guruswamy and J.A. McNeely (eds.). Protection of Global Biodiversity, Converging Strategies. Duke University Press, Durham.

Associated Press. 1997. 'Nicaraguan Rebels Surrender their Weapons.' The New York Times International News. July 22. http://www.newstimes.com/archive97/jul2297/inc.htm

Associated Press. 1993. "'My Struggle is Political, Not Economic' Says El Chacal," August 24. Cited in 'Re: Nicaragua: Hostage Crisis Update.' August 26. http://www.stile.lut.ac.uk/~gyedb/STILE/Email0002065/m6.html

Baumeister, E. 1997. Migraciones internas en Nicaragua. OIM/INEC/UNFPA, Managua.

Bendaña, A. 1991. La tragedia campesina. Editorial CEI, Managua.

Burke, Pamela. 1995. 'Native Peoples of Nicaragua.' http://www.bos.umd.edu/cidcm/mar/indnic.htm

CAPRI (Centro de Apoyo a Programas y Proyectos). 1998. Región Autónoma del Atlántico Norte, El desafio de la autonomía. Second Edition. CAPRI, Managua.

Chamorro, E. 1999. "Nos engañan, dicen YATAMA amenaza combatir al gobierno liberal." April 9:3B.

Coca Palacios, L. 1998. 'Negociaciones a fuego lento.' 7 Días. No. 154. June 18-24:23.

Comisión Nacional de Bosawás. 1995. "2da. Sesión extraordinaria, 17 y 18 de julio de 1995, Hotel Selva Negra–Matagalpa." July. Secretaría Técnica de Bosawás (SETAB), Managua.

Cuadra, E. and A. Saldomando. 1998. 'Pacificación, gobernabilidad y seguridad ciudadana,' pp. 105-38, in Orden social y gobernabilidad en Nicaragua 1990-1996 (editors Elvira Cuadra, Andrés Pérez, and Angel Saldomando). CRIES, Managua.

DIRB (Policy, Planning and Research Branch, Research Directorate, Immigration and Refugee Board of Canada). 1997. "Nicaragua Update." September. Ottawa, Canada.

Envío. 2000. Jefe del Fuac asesinado. Envío, No. 217, April: 16.

Flores, J. 1998a. "YATAMAs realizan más protestas en la RAAN." La Tribuna. May 8.

Flores, J. 1998b. "Indígena muere en refriega. YATAMAs secuestran a soldados en Bismuna." La Tribuna. May 14.
FUAC (Frente Unido Andrés Castro). 1997a. "Protocolo de Demandas Sociales del Frente Unido Andrés Castro (FUAC)." Siuna.
FUAC. 1997b. "Protocolo de Demandas del Frente Unido Andrés Castro (FUAC)." Siuna.
Gomez Nadal, P. 1998. "Acusan al gobierno de incumplir el 90% de los acuerdos de desmovilización, Al FUAC se le acabó la paciencia" 7 Días. No. 154. June 18-24:24.
González, A. 1997. "Costa: elecciones sobre arenas movedizas." Envío. No. 178-179. January-February: 23-35.
González Silva, M. 1997. 'Frente Unido Andrés Castro: We are the Protection of the Campesino.' El Nuevo Diario. August 29. http://members.aol.com/galpres/Pages/Castro.html
Government of Nicaragua. 1997. Acta de Acuerdo Final de Paz entre el Gobierno de la República de Nicaragua y el Frente Unido Andrés Castro (FUAC). Managua. December 3.
Hale, C.R. 1994. Resistance and Contradiction, Miskitu Indians and the Nicaraguan State, 1894-1987. Stanford University Press, Stanford.
Howard, S. 1997. Livelihoods, Land Rights, and Sustainable Development in Nicaragua's Bosawas Reserve. In pp. 129-141. J. De Groot and R. Ruben, eds. Sustainable Agriculture in Central America. St. Martin's Press, New York.
Hurtado, C. et al. 1992. 'Acuerdos Básicos de Puerto Cabezas.' February 23.
Leist, D. 1998. "Nicaragua Update" Central American Update. September-October. http://www.geocities.com/CapitalHill/Senate/9126/nicaragua.html
López, V. 1999. "Dicen que ahora sí se desalzaron los YATAMAs." El Nuevo Diario. February 16:2.
Nietschmann, B. 1990. Conversation by conflict in Nicaragua. Natural History. 11/90: 41-48.
Nicaragua Network Hotline. 1994. "Nicanet Hotline 02/28/94." February 28. http://www.stile.lut.ac.uk/~gyedb/STILE/Email0002062/m34.html
Nicaragua Network. 1993. "Nicanet Hotline 07/05/93." Nicaragua Network Hotline. July 5. http://www.stile.lut.ac.uk/~gyedb/STILE/Email0002062/m1.html
PACDC (Program for Arms Control, Disarmament, and Conversion). 1998. "Small Arms and Light Weapons Events/Nicaragua." PACDC, Monterey. http://pacdc.miis.edu/Central%20America/nicaragua.html
Ramirez, E., M. Ardon, and E. Holt. 1994. Diagnóstico para la identificación de acciones para un programa de desarrollo sostenible en la Reserva Bosawas Nicaragua. Commission of European Communities, Ministry of Foreign Relations of France, Managua.
SETAB (Secretaria Técnica de Bosawas). 1996. Normas generales y principios conceptuales para el uso de la tierra (ordenamiento ambiental–territorial) de la Reserva Nacional de Recursos Naturales Bosawas y su Zona de Amortiguamiento, SETAB, Managua, June.
Stocks, A. 1995. "Land Tenure, Conservation, and Native Peoples: Critical Development Issues in Nicaragua." Paper present at the Applied Anthropology Meetings, March 29 to April 2.

Stocks, A. 1998. "Indigenous and Mestizo Settlements in Nicaragua's Bosawas Reserve: The Prospects for Sustainability." Paper presented at the Annual Meetings of the Latin American Studies Association in Chicago on September 24-26.

United States Department of State. 1998. "Nicaragua Country Report on Human Rights Practices for 1997." Bureau of Democracy, Human Rights, and Labor. Washington DC, January 30.

Forests in the Time of Violence: Conservation Implications of the Colombian War

María D. Álvarez

SUMMARY. Forest remnants in the Colombian Amazon, Andes, and Chocó are the last repositories of a highly diverse and endemic biota. Historical changes in the Colombian landscape have been dramatic, but the magnitude and rate of change has increased over the last half century, while conflict has consumed the capacity of Colombian society to respond to environmental threats. Academic experts in the study of the Colombian conflict have explored the social, political, and economic implications of the war. However, the environmental consequences of conflict are documented only when groups in conflict target salient economic resources. This paper presents the first analysis of the geographic distribution of forest remnants in relation to armed conflict in Colombia. Results show that guerrillas and/or paramilitaries range throughout areas of human encroachment into remnant forests. The policies promoted by

María D. Álvarez is affiliated with Columbia University, Mail Code 5557, New York, NY 10027 (E-mail: mda2001@columbia.edu).

The author thanks O. Hernández, M.V. Llorente, C. McIlwaine, G. Martin, M. Pinedo-Vasquez, A. Reyes, M. Rodríguez, E. Sanderson, K. Willett, and the man in the Wellington boots for providing invaluable resources for this study. The author is also grateful for comments by G. Andrade, S. Price, R. Sears, A. Westing, and an anonymous reviewer which greatly improved the quality of the text.

María D. Álvarez is supported by Columbia University.

The author dedicates this paper to her grandparents–C. Álvarez and T. Sánchez–campesinos displaced by that other "Violencia" in the 1940s.

[Haworth co-indexing entry note]: "Forests in the Time of Violence: Conservation Implications of the Colombian War." Álvarez, María D. Co-published simultaneously in *Journal of Sustainable Forestry* (Food Product Press, an imprint of The Haworth Press, Inc.) Vol. 16, No. 3/4, 2003, pp. 49-70; and: *War and Tropical Forests: Conservation in Areas of Armed Conflict* (ed: Steven V. Price) Food Products Press, an imprint of The Haworth Press, Inc., 2003, pp. 49-70. Single or multiple copies of this article are available for a fee from The Haworth Document Delivery Service [1-800-HAWORTH, 9:00 a.m. - 5:00 p.m. (EST). E-mail address: getinfo@haworthpressinc.com].

Colombia's irregular armed forces range from "gunpoint conservation" rarely applied by guerrillas, to the rapid conversion of forests and crops to cattle ranches and coca (*Erythroxylum* sp.) plantations, following paramilitary occupation. Because the rates and extent of fragmentation are linked to such land use practices, armed groups may play a crucial role in determining the fate of Colombia's forests and their endemic biota.

KEYWORDS. Colombia, biodiversity, forest conservation, forest fragmentation, deforestation, forest policy, armed conflict, war, violence, economic incentives, illicit crops, coca, *Erythroxylum*, guerrillas, paramilitaries, drug trafficking

INTRODUCTION

With forests covering the largest portion of its territory–an estimated 50%–and a dizzying array of ecosystems, Colombia ranks among the select few "megadiversity" countries of the world (McNeely et al. 1990; Myers et al. 2000). Analyses that use endemism and biodiversity as criteria for setting world or continent-wide conservation priorities invariably highlight Colombian forests in their results (e.g., Olson and Dinerstein 1998). Most of these studies also include some measure of forest fragmentation threats from demographic pressure, economic development patterns, or both (e.g., Myers et al. 2000).

Forest fragmentation processes are neither recent nor uniform. In some regions, deforestation dates back to pre-Columbian times (Cavelier et al. 1998; Etter and van Wyngarden 2000). Not surprisingly, areas comprising already fragmented landscapes, principally the Andes, inter-Andean valleys, and the Caribbean region, were quickly settled by European colonizers (Etter and van Wyngarden 2000). The ensuing population distribution persists to this date, with 70% of Colombians concentrated in the three Andean cordilleras (World Wide Fund for Nature [WWF] 2000). As a result, Colombia has lost 73% of its continuous montane forests (Cavelier and Etter 1995). Large-scale expansion into the lowlands is relatively recent, and therefore the rainforests of the Chocó and Colombian Amazonia are not yet considered as threatened as the Andean forests (Bryant et al. 1997).

Nevertheless, all Colombian forests face fragmentation to some degree, as noted in the country programs of major environmental non-governmental organizations (Conservation International [CI] 2000; World Resources Institute [WRI] 2000; WWF 2000). Indeed, country wide deforestation rates in Colombia have been estimated somewhere between 0.6% (WRI 1995) and 2.8% per year (Myers 1989), with local measurements as high as 4.4% (Viña and Cavelier 1999). The critical importance of Colombian forests in the maintenance of biodiversity cannot be overstated. Estimates of species richness for well-known taxa rank Colombia as the country with the greatest number of birds, in addition to over 45,000 species of angiosperms (Stattersfield et al. 1998; McNeely et al. 1990). Many of these organisms are completely forest-dependent and endemic to specific regions of the Andes, Chocó, or western Amazonia. In effect, Colombia harbors significant parts of the Chocó and the tropical Andes, which contain 0.8% and 5.7% of the world's endemic vertebrates, and 0.8% and 6.7% of endemic angiosperms respectively (Myers et al. 2000).

International conservation programs aimed at preserving this spectacular diversity in forest remnants note that in addition to the suite of common socioeconomic challenges to conservation in developing countries, Colombia's chronic violence further jeopardizes local and international efforts to curb environmental decline (CI 2000; WWF 2000; WRI 2000). This fact is sometimes dismissed as an anomaly to be resolved by national institutions (for example, see the "Threats/Opportunities" section of WWF 2000). The Colombian war, however, dates back to the middle of the twentieth century, shows no signs of abatement, and has increased in scope and scale over the last decade.

In Colombia, the conflict has become so pervasive that the study of armed conflict within the social sciences has a name of its own, *violentología*. Because most *violentólogos* focus on socioeconomic and political issues, environmental damages have not been central in the discussion about the Colombian conflict. The socioeconomic implications of the predominant patterns of exploitation of natural resources are a perennial explanation of the origins and development of the conflict (Kalmanovitz 1994; Pizarro-Leongómez 1996). But whether the conflict per se has any environmental effect is not known. Perhaps the only recurrent element of environmental concern in the study of violence is the damage that explosives detonated along petroleum pipelines cause in local watersheds (e.g., Peñate 1991). The environmental consequences of armed conflict have received little scientific study, and except for petroleum spill damage, are unknown.

There are two main players whose actions defy the administrative and military role of the state: guerrillas and paramilitaries. The two most powerful left-wing guerrilla groups are the Fuerzas Armadas Revolucionarias de Colombia (FARC) and the Ejército de Liberación Nacional (ELN). Both groups claim to fight for communist revolutionary ideals and trace their history back to 1948 and the mid-1960s, respectively (Echandía 1999). As guerrilla activities increased in the early 1980s, wealthy landowners responded by sponsoring right-wing paramilitaries (Cubides 1999). Whether or not the paramilitaries were or are linked to the Colombian military (and thus to the government) is the subject of political and academic debate (Cubides 1999; Chernick 1998; Human Rights Watch 2000).

This paper presents results from the first analysis of the geographic distribution of forest remnants in relation to armed conflict. Two objectives are thus accomplished. The first is to estimate the proportion of forests in municipalities where guerrilla and paramilitary activities may impact resource use by individuals. It is implied that these forested areas are, to some extent, beyond the complete access and control of the state. That the Colombian government perhaps never had control over its vast forested territories (L. G. Baptiste, pers. comm.) is neither denied nor discussed because it is irrelevant to this estimation.

The second objective is to propose hypotheses about the effects of armed conflict on deforestation and forest conservation within these municipalities. Because the parties in conflict undertake policies that are directly related to natural resource use, including the preservation or exploitation of forests, some effect of armed conflict on forests can be reasonably expected. In some of the areas affected by armed conflict, governmental and non-governmental actors continue to formulate forestry policies and undertake forestry projects (Rodríguez and Ponce 1999; A. Villa, pers. comm.). This paper does not claim otherwise, rather, it assumes that armed conflict affects resource use by individuals and, based on field observations, suggests tests to detect its effects.

METHODS

Forest cover and forest types are summarized from Etter (1998), with his nomenclature in parentheses. "Andean" forests comprise sub-Andean humid (14), Andean humid (16), high-Andean humid and 'cloud' (18a), high-Andean dry (18b), Andean oak (18c), and dry and humid páramos (19-20). Tropical forests of the middle Magdalena (1m, 3m) and Caribbean forests (23) were lumped in the Andean category, and they amount to less than 1% of the total. "Chocó" forests comprise tall dense Pacific forests (2c, 3c, 7), flooded Atrato and Pacific forests (46-48a), and hyper-humid mangroves (50). "Amazon" forests comprise tall dense Amazon and Orinoco basin forests (1a, 2a, 3a, 2b, 2d, 3b, 4, 5a, 5b, 6), sub-montane, montane, and "cloud" forests of La Macarena (24-26), tall and dense Amazonian forests (42, 43), middle dense Amazonian (44), and gallery forests of the Orinoco basin (45).

Ranges of guerrillas, paramilitaries, drug crops, and land purchases by traffickers are taken from Reyes (1999). Ranges of paramilitaries reflect their presence or absence from 1985-97. Guerrilla intervention was plotted from the medium (four to ten attacks) and high activity (more than ten attacks) from 1985-97, as documented in Reyes (1999). The entire municipality is marked as affected if guerrilla actions, paramilitary activity, land purchases by traffickers, or illicit crops are detected, regardless of the actual extent of intervention. Because municipalities in the Chocó and the Amazon are very large, forests highlighted there are not proportional to the extent of armed actions, land purchases, or illicit crops in those regions (see Figure 1 for forest "types"). Since political divisions are much smaller in the Andes, the extent of these same phenomena in affected Andean municipalities will be overestimated to a lesser degree. Presence of armed groups, illicit crops, and land purchases by traffickers was overlain and measured on the forest cover map using ArcView 3.2.

Information on the management policies of guerrilla groups was obtained through conversations with local civilians and members of these armed groups in the Macarena mountains (1995), Munchique National Park and Tambito Nature Reserve (1995, 1997-99), the San Lucas mountain range (1998), and the Churumbelos mountains (1998). Additional information on environmental policies of armed actors was drawn from ELN (2001), FARC (2001), Chernick (1998), Cubides (1999), Echandía (1999), and Vargas-Meza (1998).

RESULTS: ARMED CONFLICT, ILLICIT CROPS, AND FORESTS IN COLOMBIA

Large portions of all types of forest lie in municipalities affected by armed groups, with the highest values calculated for Andean forests

FIGURE 1. Outlines of the Colombian *departamentos*, place-names mentioned in the text, and main geographic features of Colombia. Lowland forests of Chocó in light green, montane and submontane forests of the Andes in darker green, and lowland forests of the Amazon and Orinoco basin in the darkest green (see "Methods" for specific forest types). Main forested areas discussed in the text: A, Sierra Nevada de Santa Marta; B. Serranía del Perijá; C, Serranía de San Lucas (Central Andes); D, Nudo del Paramillo (West Andes); E, Serranía de Los Saltos (north), Serranía del Baudó (south); F, Páramo de Las Hermosas; and G, Serranía de La Macarena. Cordilleras of the Andes (dark green font): H, West Andes; I, Central Andes; and J, East Andes. Main rivers (blue font): K, Cauca; L, Magdalena; M, Meta; N, Caquetá; and O, Putumayo.

(Table 1, Figures 2, 3). Andean forests are also disproportionately located in municipalities experiencing conflict as inferred from the presence of guerrillas and paramilitaries (Table 1). Forests in municipalities experiencing conflict totaled 20% of the total remaining forest of Colombia (Table 1 and Figure 4). The most significant Andean forests in municipalities experiencing conflict include the Sierra Nevada de Santa Marta, the Serranía del Perijá, the Serranía de San Lucas, the Nudo del Paramillo, and the Páramo de Las Hermosas (Figures 1 and 4). The northeastern tip of the East Andes and the eastern slope of the East Andes in the departments of Meta and Putumayo also lie in municipalities experiencing conflict (Figure 4).

Chocoan forests in municipalities experiencing conflict include the western half of the Serranía de los Saltos and the lowlands of the department of Valle del Cauca. In the Colombian Amazon, the forests of the upper and middle Guaviare, Caquetá, and upper Putumayo river basins are in municipalities with ongoing conflict. The Serranía de la Macarena and the foothills of the East Andes–where some Amazon forest is found–are also experiencing conflict (Figure 4). In addition to their investments in municipalities experiencing conflict or having armed groups, traffickers have also made significant land purchases in non-conflict areas of the Chocó, the Serranía del Baudó, and lowlands of Nariño (Figure 2).

TABLE 1. Forest Types and Sociopolitical Factors

Number of hectares and percentage of total remnant forests of each forest type located in municipalities affected by armed groups, conflict, and all factors considered: guerrillas, paramilitaries, illicit crops and land purchases by traffickers. The total extent of the categories Andes, Chocó, and Amazon is 9.5, 4.3, and 35 million ha, respectively. See "Methods" for details. Percentages of total remaining forests are fractions of the estimated 48.8 million ha of forests of all types (including secondary growth) remaining in Colombia (Etter, 1998).

Forests in municipalities with	Andes		Chocó		Amazon		% of total remaining forests
	10^6 ha	%	10^6 ha	%	10^6 ha	%	
Presence of armed groups	6.63	70	1.9	44	7.48	21	33
Armed conflict (inferred)	3.42	36	1.00	23	4.74	13	18
All factors present	1.39	15	0.04	1	0.88	2	5

FIGURE 2. Forests in municipalities affected by, clockwise from top left, guerrillas (pink), paramilitaries (red), land purchase by drug business (yellow), and illicit crops (ochre). Forest color codes as in Figure 1.

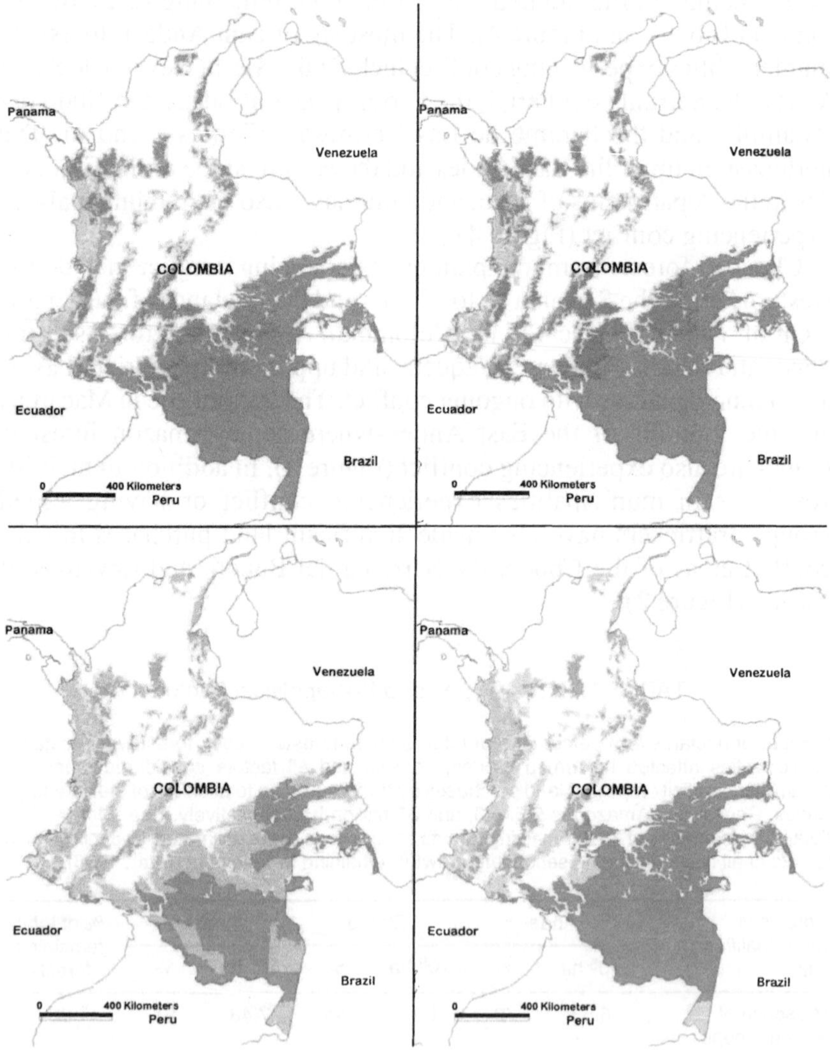

FIGURE 3. Forests in municipalities affected by non-governmental armed groups (gray). Point sites of documented "gunpoint conservation" are marked with dots, see "Methods" for sources, details. Forest color codes as in Figure 1.

DISCUSSION

Critical portions of Colombia's Andean, Chocoan, and Amazonian forests are located in municipalities where armed groups contest political authority (Figure 3). This is perhaps most evident in the case of the Andes, where the largest continuous forest remnants (Serranía de San Lucas, Nudo del Paramillo, Páramo de Las Hermosas, and East Andes)

FIGURE 4. Forests in municipalities affected by conflict between non-governmental armed groups (black). Armed conflict is inferred from the simultaneous presence of medium and high guerrilla activity and the presence of paramilitaries from Reyes (1999). See "Methods" for sources and details, forest color codes as in Figure 1.

lie in municipalities where both guerrillas and paramilitaries are present (Figure 4). The management policies and practices of these armed groups are not homogeneous in space or in time. Neither can it be assumed that their policies and practices are applied throughout their respective ranges or "territories." The following sections propose hypotheses about the scope and trends of environment-related actions by armed groups in order to characterize the effects of armed conflict on Colombia's forests.

This study examines three processes whereby armed conflict may affect forest use and conservation. The first, "gunpoint conservation" involves the exclusion of most productive activities from certain areas, but seems to be a rare phenomenon. It is enforced with land mines and civilian curfews. The second involves forest fragmentation pressures from illicit crop growers and cattle ranchers in areas that are beyond the rule of law and/or contested by armed groups. The third is a consequence of development under the "violence syndrome," which is defined as the collapse of the institutional framework for civilian law and order.

Gunpoint Conservation

In a number of guerrilla-controlled areas, including parts of the Serranía de San Lucas, conservation is carried out by means of armed coercion. The ELN protects some forests in the Serranía de San Lucas purportedly because of their role in the local hydrology. Their methods include placing landmines or posting signs that warn of landmines in patches of montane forests. In addition, these forests–*el monte*–have served as refuge from air surveillance by government forces (Dávalos 2001). The control exerted by the ELN over these resources is also a demonstration of authority and a manifestation of a broader ideological program. This approach to forest preservation can be labeled "gunpoint conservation" because it is strictly enforced and maintained with the threat of violence (Figure 3). There is no evidence that the local communities would respect the ELN's conservation policies in the absence of coercion (Dávalos 2001).

The guerrilla groups have appropriated the current discourse of biodiversity and conservation, incorporating it into their revolutionary rhetoric with nationalistic undertones. On their Internet web pages, both FARC and ELN tout their environmental concerns, emphasizing sovereignty over biodiversity (FARC 2001; ELN 2001). Sometimes these interests and management initiatives coincide with the government's conservation objectives. For example, during the 1997 El Niño, campesinos seeking to expand their lots burned the lowlands of the Munchique National Park. The FARC, advised by the park's low-ranking officials, threatened those responsible for the fires. Consequently, and despite the continuing drought, the fires stopped in late August.

However, there may be other considerations motivating the application of these policies. For example, as of 1995, the FARC excluded almost all agriculture and hunting from the southern half of the Serranía de la Macarena. They claimed to preserve the wealth and beauty of the forests for the benefit of future generations, but these forests also housed their national headquarters at the time. The guerrillas construe the forest as shelter from air raids (strategic cover), an invaluable source of water (reservoir), or a necessary condition for the protection wildlife and other biodiversity values (natural habitat).

In the latter interpretation, some charismatic animals have an intrinsic value that the guerrillas are willing to protect. Interestingly, these values are more attached to the animals than to the forest itself. In the Serranía de San Lucas, FARC agents expressed their preoccupation regarding the plight of a spectacled bear (*Tremarctos ornatus*) whose partner in captivity had died. They wanted either to build a zoo for the local villagers or send the lonely male to an urban facility that could care for him. This may partly explain the zeal with which the guerrillas preserve some forest fragments, but their practical and strategic considerations provide a more parsimonious explanation for their conservation initiatives.

These policies, however, are not static or evenly applied. In August 2000, the Colombian newspaper *El Tiempo* reported the construction of more than 200 km of roads linking the demilitarized zone (DMZ)–also known as the Military Exclusionary Zone–in southeast Colombia to the southern outskirts of Bogotá through the Macarena, Tinigua, Picachos, and Sumapaz National Parks. The following month, television footage confirmed the road construction. According to journalistic versions, the purpose of the road would be to facilitate the entry of hostages and reagents for drug processing into the DMZ. The construction of the road seems to have paralleled an unprecedented growth in illicit agriculture in the DMZ, where coca (*Erythroxylum* sp.) plantations increased 32% over the 1999-2000 period (Maserié 2001). However, it is not clear whether the new illicit crops within the DMZ replaced agricultural crops or natural habitats. The construction of this road suggests that economic interests override the environmental agenda that the FARC had claimed to promote by, for example, including the sustainable development of natural resources as an item in the agenda for peace negotiations with the government (Ricardo et al. 2000).

Peace negotiations with the ELN are moving towards establishing a second DMZ in the Serranía de San Lucas. For updates on this process, the reader is referred to the Internet web pages of the Presidencia de la República and Proceso de Paz (http://www.presidencia.gov.co/webpresi/ and http://www.procesodepaz.com/). Programs for the protection of river basins, top soil, flora, and fauna have been included as one of many items in their charter, the *Reglamento para la Zona de Encuentro* (Presidencia de la República 2001). Whether or not these proposals result in environmental protection remains to be seen, particularly in the face of the government's plan to develop gold mining in the region beyond its current artisanal scale (Londoño 2001; Villaruel et al. 2000).

GUNPOINT FRAGMENTATION AND ECONOMIC INCENTIVES

The second and perhaps more pervasive type of impacts on the forest stems from the pattern of economic development that prevails in many guerrilla and paramilitary-dominated areas. This is necessarily linked to the other half of the violence equation: illicit trafficking and displacement. Because of the FARC's geographic range during the coca boom of the 1980s (Echandía 1999), this was the first politically oriented armed group to profit from illicit crops. In effect, the FARC regulate trade between growers and buyers by collecting taxes and requiring campesinos to grow three hectares of food for every hectare of coca (A. Reyes, unpublished). Paramilitaries battle for control of the land and transport routes throughout guerrilla territories. This is most efficiently accomplished by undermining their own income base: the campesinos (Echandía 1999; Cubides 1999; Chernick 2000). This policy would explain why the distribution of guerrillas and paramilitaries overlaps significantly, as shown by a comparison between their full ranges (Figure 2) and areas of overlap (Figure 4).

Because the paramilitaries are mercenaries for cattle ranchers and traffickers, landholdings usurped from 'guerrilla territories' are quickly consolidated following the displacement of the local campesinos. These so-called territories are often small holdings where political power is partially or totally administered by guerrillas, sometimes by sheer force. This would explain why the percentage of farms in Colombia over 500 ha has grown from 33% in 1984 to 50% in 1997 (Rincón 1997). The conversion of the last forest remnants to cattle ranches often follows territorial gains made by the paramilitaries. The paramilitaries have been able to evict guerrillas completely from only some parts of the Guaviare river basin and the lowlands of the department of Nariño (compare Figures 2 and 4). Their territorial conquests, however, are expected to increase as conflict escalates and the government strategically focuses its military-backed campaign to eradicate illicit crops on the southern part of the country (Chernick 1998; Vargas-Meza 1998; Cubides 1999; Tate 2000).

The environmental effects of these contests for land are not limited to the areas seized by cattle ranchers. Illicit agriculture per se has already been identified as a notorious factor in forest degradation (Cavelier and Etter 1995; Henkel 1995; Young 1996). In 1998, it accounted for up to half the annual deforestation (Álvarez, in press). Illicit crop eradication programs undertaken by the Colombian government over the last decade may have led to greater conversion pressure on the surrounding forests as growers seek new lands to make up for their lost revenue (Henkel 1995; Young 1996; Kaimowitz 1997; Reyes 1999; but see Cavelier and Etter 1995).

If the United Nations Office for Drug Control and Crime Prevention (UN-ODCP) is correct, an average hectare of coca in Colombia yields 1,630 kg of coca leaf (no time period given, presumably per year) (UN-ODCP 1999, 44). If the price of coca leaf at the farm is comparable to that found in Bolivia and Peru, US$1.44 (no prices given for Colombia, UN-ODCP 1999, 80), then an average campesino would receive US$2,347 per hectare planted with coca. This amount of coca leaf results in 3.5 kg of cocaine (see Table 11, UN-ODCP 1999, 44). It would fetch an average wholesale price of US$79,590 in the US, the main importer of Colombian cocaine (UN-ODCP 1999, 164). Thus, the campesino receives 2.9% of the wholesale price paid for the cocaine produced from his/her crop, while assuming much of the risk associated with armed conflict and law enforcement against illicit crops. Ironically, the paramilitary expansion that threatens the lives and property of coca growers is partly funded by landed traffickers.

These facts indicate that forested areas near illicit crops face at least three kinds of fragmentation pressure. First, the economic incentives for growing illicit crops are high, with expected revenue from a single hectare of coca (see above) exceeding the average per capita income of Colombia (at US$2,168) (WRI 2000). Therefore, campesinos may choose to expand their crop areas for simple economic reasons. Second, government eradication of illicit crops may lead growers to move deeper into the forests to avoid detection (Henkel 1995; Young 1996; Kaimowitz 1997; Reyes 1999). Finally, paramilitaries may attack, facilitating the expansion of cattle ranches and/or the consolidation of lands held by traffickers (Cubides 1999; Reyes 1999). If these pressures exist in rural Colombia, fragmentation threats are greater than 'background' levels found in forested areas adjacent to coca and opium poppy (*Papaver somniferum*) production (Figure 2). Here the term 'background' applies to those areas without armed conflict and illicit crops. Pressure on areas with armed conflict (Figure 4) is also expected to be higher than in background areas because the conflict diminishes local incentives for long-term management, as explained below.

These processes that transform land use and tenure may explain why forest clearing continues despite rural flight in the Chocoan lowlands of the department of Valle del Cauca (Figure 1) (Reyes 1999). By itself, the persistence of deforestation is surprising because the Colombian countryside is more sparsely populated and less economically productive today than it was ten years ago. This defies the conventional connection between population growth, economic growth, and deforestation. Between 1990 and 1995, the rural population of Colombia decreased 0.3% each year, while the total national population grew at an annual rate of 1.7% (WRI 1997). A significant part of this trend can be attributed to forced migration: 300,000 people were displaced by the conflict in 1998 alone (Reyes, 1999), for a total of 1,800,000 campesinos displaced as of 1999 (Chernick 2000).

Arguably, forest fragmentation driven by illicit trade would occur regardless of the contested status of these forested areas. Although this may be true, there is no actual way of testing the effects of illicit crops and armed groups independently since the association between them dates back to the 1980s, and state repression guarantees a lack of transparency. The paradoxical conclusion is that due to the conflict, rural population growth, and/or economic output are not directly proportional to deforestation. This provides a rebuttal to the usual malthusian description of deforestation pressures in tropical countries.

The Violence Syndrome

Are incentives for conservation altered by the war? The third type of effects involves unsustainable development not directly associated with illicit crops. Licit economic activities in violently disputed areas are expected to produce greater environmental damages *because* of the conflict (Figure 4). This can be more easily visualized through a concrete example. Because the ELN impedes the entry of investors to the vast gold deposits of the Serranía de San Lucas, gold extraction there is primitive and inefficient. At the same time, because the area is beyond the full scope of action by governmental and non-governmental environmentalists, pollution from artisanal mining is not regulated. The quick profits provided by artisanal mining attract many immigrants who exert pressure on the area's forest with hunting, settlement, and agricultural activities (Dávalos 2001).

Because these settlers live on the brink of eviction (or violent death), high economic stakes are combined with disincentives for long-term conservation or management of resources. As rational economic actors, individual artisanal gold miners forgo any measure that would limit their production, including pollution control and investment in 'cleaner' mining technology *because* the war makes their future uncertain, and the opportunity costs of not exploiting the gold are very high.

The agrarian economy as the first victim of violence. At the national scale, the prevalent agrarian development model, imposed partly by armed groups in the 1990s, tends to replace annual food crops with less productive cattle pastures (Vásquez 1997). By failing to support the institutions that assisted small farmers, most of which were downgraded or dismantled in the early 1990s, government policies have further entrenched this pattern (Robledo 1999). As the flight of the rural population to urban centers grows, and food imports replace forgone agricultural production (from 800,000 tons in 1990 to 7,000,000 tons in 1999), the expansion of cattle ranching continues throughout the agricultural frontier (Vásquez 1997). Though most of the expansion is taking place on agricultural land, at least a fraction corresponds to forest conversion.

Other things being equal, the impact of violence, estimated at 33% of agriculture's contribution to the Gross National Product (GNP) in 1995, is a staggering blow to the economy of Colombia (Bejarano et al. 1997). Ultimately, the economic adjustments that the Pastrana administration undertook in 1998, including a significant reduction in the budget for the Ministry of the Environment, can be ascribed to an economic recession that is related in important ways to the war. The decentralization of the National Natural Park System, undertaken even before these budget cuts took place, might increase regional investment in conservation (Rodríguez 1998). However, a significant part of this regional investment hinges on channeling resources from power-generating companies (Rodríguez and Ponce 1999).

As a result, more developed regions where infrastructure is concentrated receive more funds for conservation than the undeveloped areas, such as the Serranía de San Lucas or most of southeast Colombia where the threats of illicit crops and armed conflict are concentrated (Figures 1 and 2). In this framework, the current guerrilla policy of attacking the national power infrastructure represents an additional, unforeseen, armed threat to conservation (see Proceso de Paz [http://www.procesodepaz.com]). While resources from international funding agencies are also expected to support the new decentralized system (Rodríguez and Ponce 1999), security considerations will always influence the allocation of scarce funds. It is then no wonder that violence is considered the most important obstacle to the economic, social, and political development of Colombia (World Bank 1999).

This last set of effects is the most difficult to trace and outline into a valid test. In principle, the environmental consequences of this 'violence syndrome' are not different from the challenges that conservation faces in other developing countries. I contend, however, that the erosion of civilian rule and the almost absolute lack of justice adjudication in the Colombian countryside are distinct, if diffuse, threats to conservation. In terms of investment in development, individuals and companies alike cannot be held accountable for their actions if the rule of law is in shambles, as it is in rural Colombia. The flight of millions of campesinos from rural violence has left the areas affected by conflict less economically productive. Displacement also precludes the practice of forest management, restoration, or conservation.

Future studies on the relationship between armed conflict and forests should focus on quantifying localized deforestation in areas highlighted in Figures 2-4, and they should distinguish between fragmentation from illicit agriculture and that caused by the expansion of cattle pasture. When and if these data become available, the links between paramilitaries, guerrillas, cattle ranching, deforestation, and demographic changes will be established and quantified. Studies of the "violence syndrome" should compare areas of similar demographic characteristics that are with and without conflict to identify the impact that this sociopolitical phenomenon may have on forests.

RECOMMENDATIONS: WHENCE CONSERVATION?

There are no obvious ways to mitigate the effects of conflict on the forests of Colombia because the effects outlined here are ambiguous. Only some of the armed groups practice a form of forest conservation, albeit in a highly localized and coercive manner. On the other hand, the government cannot be considered fully enabled in the 33% of Colombia's forests that are found in municipalities where armed groups operate. Some of its policies conducted amidst the conflict, including illicit crop eradication and road infrastructure development, may have deleterious impacts on the forest. Despite these limitations, there are a number of policy options directly available to counter deforestation.

First, economic incentives for intensive agriculture should be restored in order to enhance rural productivity. This step is necessary to create profitable alternatives to cattle ranching, an activity that is both less productive and labor intensive than traditional agricultural systems (Robledo 1999). Increased agricultural productivity could also help counteract existing incentives for illicit crop cultivation. Most of the technical and financial support for traditional agriculture was dismantled or downgraded during the early 1990s. Even in those areas where illicit crops are not the only profitable economic alternative, agricultural production has plummeted because most campesinos have no access to technology or markets for their products. In the meantime, food imports have skyrocketed and campesinos are cornered in the crossfire between guerrillas, paramilitaries, and the government's campaign against illicit crops. Low agricultural productivity means that more land is cleared for agriculture, further depleting the forest cover, and impoverishing the natural heritage of Colombia and the world.

Second, the spraying of herbicides on coca and poppy plots in Colombian forests must cease. This technique for combating illicit crop cultivation has been ineffective in Colombia. Since 1986, the total area under illicit crop cultivation has more than quadrupled despite an 80-fold increase in the area fumigated with herbicides (UN-ODCCP 1999). The fumigation of illicit crops is also believed to create additional pressure on primary forests of Latin America by driving illicit growers deeper into the frontiers to avoid law enforcement (Henkel 1995; Young 1996; Kaimowitz 1997). Moreover, there is no indication from data on the price and overall availability of illicit drugs that this policy is achieving anything more than social upheaval and deforestation in the Colombian countryside (UN-ODCCP 1999). Indeed, the policy of persecuting poor growers who earn less than 5% of the revenue generated by illicit drugs could be questioned on humanitarian grounds alone. Because the Colombian government is bound by international agreements to reduce illicit crop production, policies for achieving this goal need to be evaluated in light of the progress made by consumer countries as well. International organizations, such as the United Nations Office for Drug Control and Crime Prevention could play an instrumental role in investigating alternative policies (e.g., market approaches) for eliminating illicit crops.

Third, international and Colombian environmental agencies need to prepare for enforcement of conservation in those areas that are currently under a "gunpoint conservation" regime. While these areas represent a minuscule fraction of the "guerrilla" areas (Figure 2), their protection would cease with the return of guerrillas into civilian life. With peace negotiations in place or underway, the possibility of full-blown, large-scale, unplanned exploitation becomes certain in areas that are now off-limits from security risks. Some areas of exceptional biological importance, like Munchique National Park, the Serranía de la Macarena, and the Serranía de San Lucas, should receive special protection against such future threats. Not only is the future of 10% of the world's biodiversity at stake (McNeely et al. 1990), but the stability of the watersheds that support many Colombian cities–a more politically salient justification to preserve forests–will be compromised if strong protective measures are not taken soon.

The measures described above are paltry steps when compared to the efforts required by the more fundamental challenge that Colombian society faces: the reconstruction of the nation and the validation of its own legitimacy. This reconstruction will necessarily involve the country's forest and biodiversity resources. Perhaps then the lessons learned from decades of violence will come to fruition in national policy.

REFERENCES

Álvarez, M.D. in press. Illicit crops and bird conservation priorities in Colombia. Conservation Biology.

Bejarano, J.A., C. Echandía, R. Escobedo and E. León. 1997. Colombia: inseguridad, violencia y desempeño económico en las áreas rurales. Fondo Financiero de proyectos de Desarrollo-Universidad Externado de Colombia, Bogotá.

Bryant, D., D. Nielsen, and L. Tangley. 1997. The last frontier forests: ecosystems and economies on the edge: what is the status of the world's remaining large, natural forest ecosystems?. World Resources Institute, Forest Frontiers Initiative, Washington, DC.

Cavelier, J. and A. Etter. 1995. Deforestation of montane forest in Colombia as result of illegal plantations of opium (*Papaver somniferum*). Pp. 541-549 in P. Churchill, H. Baslev, E. Forero, and J.L Luteyn (eds). Biodiversity and conservation of neotropical montane forests. New York Botanical Garden, Bronx, New York.

Cavelier, J., T.M. Aide, C. Santos, A.M. Eusse, and J.M. Dupuy. 1998. The savannization of moist forests in the Sierra Nevada de Santa Marta, Colombia. Journal of Biogeography, no. 25:901-912.

Chernick, M.W. 1998. The paramilitarization of the war in Colombia. NACLA Report on the Americas 31(5):28-33.

Chernick, M.W. 2000. Elusive peace: Struggling against the logic of violence. NACLA Report on the Americas 34 (2):34-37.

CI (Conservation International). (2000, July 13). Colombia Program Extended Overview [Overview]. Washington, DC. Conservation International Foundation. Retrieved March 16, 2001 from the World Wide Web: http://www.conservation.org/web/fieldact/regions/TTAREG/colomb1.htm

Cubides, F. 1999. Los paramilitares y su estrategia. pp. 151-199 in Llorente, M.V. and M. Deas (eds). Reconocer la guerra para construir la paz. Editorial Norma, Bogotá.

Dávalos, L.M. 2001. The San Lucas mountain range in Colombia: how much conservation is owed to the violence? Biodiversity and Conservation, no. 10: 69-78.

Echandía, C. 1999. Expansión territorial de las guerrillas colombianas: geografía, economía y violencia. pp. 99-149 in Llorente, M.V. and M. Deas (eds). Reconocer la guerra para construir la paz. Editorial Norma, Bogotá.

ELN (Ejército de Liberación Nacional). 2001. Recursos Naturales [Announcement]. Unknown place: Ejército de Liberación Nacional. Retrieved 16 March, 2001 from the World Wide Web: http://www.eln-voces.com/f_todo_recursos.htm

Etter, A. and W. van Wyngarden. 2000. Patterns of landscape transformation in Colombia, with emphasis in the Andean region. Ambio 27 (7):432-439.

Etter, A. 1998. Mapa general de ecosistemas de Colombia. In: Informe Nacional sobre el Estado de le Biodiversidad 1997 Colombia. Instituto de Investigación de Recursos Biológicos Alejandro von Humboldt, PNUMA, Ministerio del Medio Ambiente, Bogotá.

FARC (Fuerzas Armadas Revolucionarias de Colombia). 2001. La Amazonía objetivo del Imperio [Article]. Unknown Place: Fuerzas Armadas Revolucionarias de Colombia. 13 March. Retrieved March 16, 2001 from the World Wide Web: http://www.farc-ep.org/

Henkel, R. 1995. Coca (*Erythroxylum coca*) cultivation, cocaine production, and biodiversity loss in the Chapare region of Bolivia. pp. 551-560 in Churchill, S.P., H. Balslev, E. Forero, and J.L. Luteyn (eds). Biodiversity and conservation of neotropical montane forests. The New York Botanical Garden, Bronx, New York.

Human Rights Watch. 2000. COLOMBIA: The Ties That Bind: Colombia and Military-Paramilitary Links. Reports 12 (1B). [Report]. Human Rights Watch, New York. Retrieved March 16, 2001 from the World Wide Web: http://www.hrw.org/reports/2000/colombia/

Kaimowitz, D. 1997. Factors determining low deforestation: the Bolivian Amazon. Ambio 26 (8):536-540.

Kalmanovitz, S. 1994. Economía y nación: una breve historia de Colombia. CINEP, Universidad Nacional, Bogotá.

Londoño, J.G. 2001. Colombia presenta nuevo mapa minero: Yacimientos de oro en zona de guerra. El Tiempo. March 12.

Maserié, S.G. 2001. Colombia será certificada hoy: La coca crece 32% en zona de despeje. El Tiempo. March 1.

McNeely, J.A., K.R. Miller, W.V. Reid, R.A. Mittermeier, and T.B. Werner. 1990. Conserving the world's biological diversity. The World Conservation Union, World Resources Institute, Conservation International, World Wildlife Fund, and The World Bank, Gland, Switzerland.

Myers, N. 1989. Deforestation rates in tropical forests and their climatic implications. Friends of the Earth, London.

Myers, N., R.A. Mittermeier, C.G. Mittermeier, G.A.B. da Fonseca, and J. Kent. 2000. Biodiversity hotspots for conservation priorities. Nature 403: 853-858.

Olson, D. and E. Dinerstein. 1998. The Global 200: A Representation Approach to Conserving the Earth's Most Biologically Valuable Ecoregions. Conservation Biology 12(3):502-515.

Peñate, Andres. 1991. Arauca: politics and oil in a Colombian province. M.Sc. Thesis. St. Anthony's College-Oxford University, Oxford (UK). Cited in Echandía, C. 1999. Expansión territorial de las guerrillas colombianas: geografía, economía y violencia. pp. 99-149 in Llorente, M.V. and M. Deas (eds). Reconocer la guerra para construir la paz. Editorial Norma, Bogotá.

Pizarro-Leongómez, E. 1996. Insurgencia sin revolución: la guerrilla colombiana en una perspectiva comparada. Tercer Mundo Editores, Bogotá.

Presidencia de la Republica. 2001. Texto completo del reglamento para la zona de encuentro ELN [Announcement]. Bogotá: Gobierno Nacional de Colombia and ELN. Retrieved March 16, 2001 from the World Wide Web: http://www.presidencia.gov.co/eln/index.htm

Reyes, A. 1999. Especial: 35 años de conflicto. Lecturas Dominicales. El Tiempo, 17 October.

Ricardo, V.G., F. Valencia-Cossio, M.E. Mejía, N. Restrepo, R. Espinosa, R. Reyes, J. Gómez, and F. Ramírez. 2000. Agenda común por el cambio hacia una nueva Colombia. [Agenda]. Colombia: Gobierno Nacional de Colombia and FARC, 6 May. Retrieved March 16, 2001 from the World Wide Web: http://www.dialogos.com.co/

Rincón, C. 1997. Estructura de la propiedad rural y mercado de tierras. Postgraduate Thesis. Universidad Nacional de Colombia, Bogotá.

Robledo-Castillo, J.E. 1999. Neoliberalismo y desastre agropecuario. Deslinde, no. 25:32-49.

Rodríguez, M. 1998. La reforma ambiental en Colombia: anotaciones para la historia de la gestión pública ambiental. Tercer Mundo Editores, Bogotá.

Rodríguez, M., and E. Ponce. 1999. Financing the green plan ('Plan Verde') in Colombia: challenges and opportunities. Paper presented at the Workshop on Financing of sustainable forest management PROFOR, UNDP Programme on Forests, 11-13 October 1999, London.

Stattersfield, A.J., M.J. Crosby, A.J. Long, and D.C. Wege. 1998. Endemic Bird Areas of the World: Priorities for Biodiversity Conservation. BirdLife Conservation Series No. 7, BirdLife International, Cambridge, UK.

Tate, W. 2000. Repeating past mistakes: aiding counterinsurgency in Colombia. NACLA Report on the Americas 34 (2):16-19.

UN-ODCCP (United Nations Office for Drug Control and Crime Prevention). 1999. Global illicit drug trends. UN Publication Series No. E 99 XI 15. United Nations, New York.

Vargas-Meza, R. 1998. A military-paramilitary alliance besieges Colombia. NACLA Report on the Americas 32 (3):25-27.

Vásquez-Ordoñez, R. La agricultura colombiana en 1996. Agronomía Colombiana XIV (2):158-181.

Villaruel. J., J.H. Ochoa, J.M. Molina, L. Alvarado, J.L. Navarro, L. Bernal, L.E. Jaramillo, R. Salinas, C. Sánchez, H. Castro, and J. Buenaventura. 2000. Minerales estratégicos para el desarrollo de Colombia. UPME, Minercol, Ingeominas, Bogotá.

Viña, A. and J. Cavelier. 1999. Deforestation rates (1938-1988) of tropical lowland forests on the Andean foothills of Colombia. Biotrópica 31(1):31-36.

World Bank. 1999. Violence in Colombia: building sustainable peace and social capital. Report No. 1865-CO, The World Bank, Washington, DC.

WRI (World Resources Institute). 1995. World resources 1994-1995: a guide to the global environment. World Resources Institute, Washington, DC.

WRI. 1997. World resources 1996-1997: a guide to the global environment. World Resources Institute, Washington, DC.

WRI. 2000. Facts and figures: country environmental data. World Resources Institute, Washington, DC.

WWF (World Wide Fund for Nature). 2000. WWF's Latin America & Caribbean programme: Colombia [Program]. World Wide Fund for Nature, Washington, DC. Retrieved March 16, 2001 from the World Wide Web: http://www.farc-ep.org/

Young, K.R. 1996. Threats to biological diversity caused by coca/cocaine deforestation. Environmental Conservation 23(1):7-15.

Lessons Learned from On-the-Ground Conservation in Rwanda and the Democratic Republic of the Congo

Andrew J. Plumptre

SUMMARY. The war and genocide that swept through Rwanda and the eastern Democratic Republic of the Congo has severely disrupted conservation activities there. However, those conservation organizations that have stayed in the region have been able to achieve a considerable amount. During 1998, I carried out a survey of staff from two conservation projects in Rwanda to determine what had motivated them to carry on working despite the loss of all senior staff, the suspension of regular salaries, and threats to their lives. The two projects were the Nyungwe Forest Conservation Project supported by the Wildlife Conservation Society (WCS) and the Karisoke Research Center supported by the Dian Fossey Gorilla Fund International. Staff at both sites felt that continuing to work had increased the risk to their lives. The factors that motivated staff to continue working included their expectation of future payment of salaries, the dedication of conservation organizations to supporting the

---

Andrew J. Plumptre is Director, Albertine Rift Program, Wildlife Conservation Society, Kampala, Uganda.

The author is very grateful to Jean-Bosco Bizumuremyi and Felix Mulindahabi for carrying out the interviews and to Michel Masozera and Liz Williamson for helping supervise the work in the field. The author thanks Bill Weber and Amy Vedder for kindly commenting on earlier drafts.

Support for the research presented here came from WCS, Karisoke Research Center, and Office Rwandais pour la Tourisme et Parcs Nationaux.

[Haworth co-indexing entry note]: "Lessons Learned from On-the-Ground Conservation in Rwanda and the Democratic Republic of the Congo." Plumptre, Andrew J. Co-published simultaneously in *Journal of Sustainable Forestry* (Food Product Press, an imprint of The Haworth Press, Inc.) Vol. 16, No. 3/4, 2003, pp. 71-91; and: *War and Tropical Forests: Conservation in Areas of Armed Conflict* (ed: Steven V. Price) Food Products Press, an imprint of The Haworth Press, Inc., 2003, pp. 71-91. Single or multiple copies of this article are available for a fee from The Haworth Document Delivery Service [1-800-HAWORTH, 9:00 a.m. - 5:00 p.m. (EST). E-mail address: getinfo@haworthpressinc.com].

sites, and the dedication of senior field staff who were forced to flee. Notably, field staff continued working because they felt that they were protecting an important part of their natural heritage, and they believed their work was important for their country. The survey also asked staff what the conservation organizations could do better in future war situations. One of the major lessons learned from the study was the important role that junior staff members play in holding such projects together during armed crisis. Most conservation organizations focus on training the senior staff but do not create a career structure for the junior staff. This finding has led WCS to plan more training and give more responsibilities to junior staff members.

KEYWORDS. Civil war, Rwanda, Democratic Republic of the Congo, lessons learned, motivations of staff

I made a mistake. War was predictable and I should have planned for it.

–Mike Fay, WCS, 1997

INTRODUCTION

Over the last ten years, civil war has hindered efforts to conserve Rwanda's wildlife and protected areas. The war that started in October 1990, the genocide that followed in 1994, and the general state of insecurity that lasted until 1999 have made work and tourism in the forests of Rwanda very hazardous. Dozens of staff members from the country's parks and forest reserves have become victims of the violence. The decline in international tourism virtually eliminated an important source of income to the Rwandan Office of Tourism and National Parks (ORTPN), making it difficult for the department to function. Before the war, tourism was Rwanda's third largest source of foreign currency, bringing in about US$2 million in 1989. Likewise, since 1996, civil war in the neighboring Democratic Republic of the Congo (DRC) has severely hampered the efforts of conservation organizations and the Congolese Institute for the Conservation of Nature (ICCN). Despite the problems in these countries, the Rwandan and Congolese nationals employed on conservation projects achieved a great deal in many protected areas–often at great risk to their personal safety. Where nationals maintained a presence, they have helped to ensure the survival of protected areas by educating and lobbying incoming military personnel and political leaders. Whenever possible, they have maintained patrols in the forest to remove snares and otherwise limit poaching. The protected areas where a presence was maintained have fared significantly better than those left without staff.

Given the persistent occurrence of war and insecurity in many parts of the world, it is important to learn how best to cope with conflict and still achieve some conservation objectives. While managing field programs for the Wildlife Conservation Society (WCS) in Rwanda and the Democratic Republic of the Congo (DRC), I became interested in drawing lessons from experiences of conservationists during the civil unrest. To understand how international and local conservation capacities were affected, I analyzed the experience of WCS and, even more importantly, the experiences of Rwandan and Congolese conservationist practitioners in the field. What was important to the people trying to conserve wildlife on-the-ground during a war situation? What motivated them to continue working despite the risk to their lives? What could conservation organizations such as WCS do better in future in a similar situation? How can conservation organizations prepare for possible periods of insecurity? This paper examines the wars in Rwanda (1990-2000) and in the DRC (1996-2000) and considers lessons learned that could help improve project-level planning and management in future times of insecurity (Figure 1). Planning for the event of war can take place at various levels, from the leadership of countries and militaries to the field staff that implement conservation projects.

International agencies and governments–Conservation organizations can prepare for war by educating governments, militaries, and international relief agencies about the importance of protected areas. The United Nations High Commissioner for Refugees (UNHCR) was severely criticized for allowing refugees from the Rwandan genocide to settle near the Virunga National Park. Their presence eventually led to the deforestation of about 100 km^2 (Henquin and Blondel 1996). To help prevent such problems, the Biodiversity Support Program, supported by United States Agency for International Development (USAID), has been working with relief agencies to encourage them to incorporate environmental considerations into relief operations and the siting of refugee camps. Similarly, lobbying efforts on behalf of protected areas may encourage governments to think twice before allowing their soldiers to feed themselves with wildlife from protected areas.

FIGURE 1. Map of the region showing the location of the protected areas mentioned in the text. The Albertine Rift runs down the border between the DRC and the neighboring countries. The Virunga Volcanoes straddle Rwanda, the DRC, and Uganda, and include three national parks (see text). The stippled portion of the Akagera National Park represents the current park area. It is the area that remained after two-thirds of the original park area was removed from protection following the genocide and civil war.

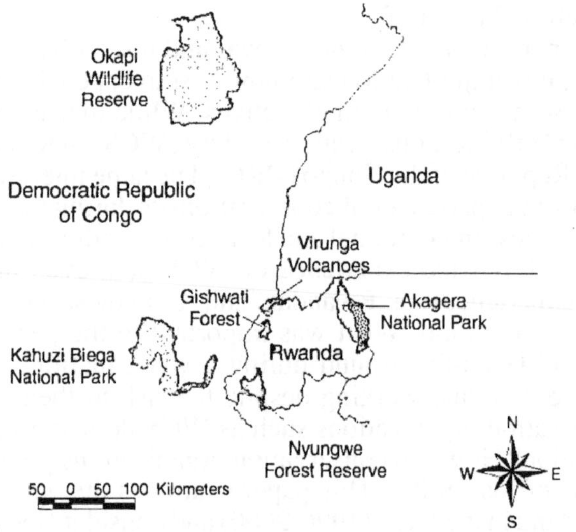

Conservation organization–Individual conservation organizations can also prepare for the event of insecurity or war by producing a security plan. This plan should have a decision making process that allows the organization to evaluate security concerns and decide whether to stay fully operational or withdraw its staff.

Individual projects–Each project should also have a security plan. The plan should include: (1) a system for information flow about the security situation; (2) plans for evacuation; (3) behaviors that should be adopted by the staff in the event of insecurity; and (4) a clear definition of the chain of authority in case someone is injured, killed, taken hostage, or has to flee.

BRIEF HISTORY OF THE CIVIL WARS

In order to understand the events that took place in the protected areas of Rwanda and DRC, it is useful to review some of the main events that took place in these countries during the fighting. The following pages provide a brief overview of what happened.

Rwanda

In 1990, the Rwanda Patriotic Front (RPF) attacked Rwanda from Uganda and established a base in the Virunga Volcanoes region. The volcanoes straddle the borders of Uganda, Rwanda, and the DRC and encompass their three national parks. The RPF was mostly composed of descendants of ethnic Tutsis who had fled Rwanda in the 1960s to avoid inter-ethnic and political violence with Hutus. They maintained a presence in the Virunga region from 1990-94. During this time, there were several international attempts to promote peace, and in August 1993 a peace accord was signed. However, a group of extremist Hutus in Rwanda were not satisfied with the peace plans, particularly the power that the RPF would have held in the armed forces.

In April 1994, the presidents of Rwanda and Burundi were killed when their plane was shot down upon returning to Rwanda from Arusha (Tanzania), where the president of Rwanda had signed an accord for power sharing with the RPF. This was immediately followed by widespread killings, leading to the genocide of an estimated 800,000 Tutsis and moderate Hutus. These killings were undertaken by Hutu militias called the *interahamwe* (literally 'those who work together') that were allied with the government. The RPF launched a major offensive to try to save the people who were being massacred, and by early July it had taken the capital, Kigali. In late June 1994, French military forces launched Operation Turquoise in southwest Rwanda to prevent further killings by either side. Some argue that this was done to protect the Rwandan government leaders that France had been supporting against the RPF (Berry and Berry 1999; Prunier 1995).

By July 17, 1994, the RPF had taken control of the rest of the country and declared an end to the war. However, during July and August, between 1.7 and 2 million Hutus fled Rwanda in fear of reprisals. To house them, massive refugee camps were established in the eastern DRC and western Tanzania. From these camps, the *interahamwe* and activists from the previous government launched guerilla attacks on civilians and military targets within Rwanda. Relative calm returned to Rwanda in 1995 and continued during most of 1996. However, sporadic attacks came from militias that were often based in the forests of protected areas. In October 1996, forces of the new Rwandan government (now the Rwandan Patriotic Army–RPA) launched an offensive with support from Uganda. It aimed to remove the *interahamwe* from the refugee camps and repatriate those who were willing to return to Rwanda but fearful to do so because of threats from the *interahamwe*. There were probably secondary reasons for the invasion, such as gaining access to the gold and diamond riches found in the DRC. This offensive was very successful and progressed much further than anyone could have believed. Eventually, it toppled President Mobutu Sese Seko, a corrupt dictator who had ruled Zaire (as the DRC was called at that time) for over 30 years.

After the return of many of the refugees, individual attacks within Rwanda escalated. The attacks were concentrated in the northwest section of the country, a political stronghold of the previous government. Refugees that had returned with smuggled arms perpetrated many of the attacks. During 1997-98, killings by the *interahamwe* and the RPF were widespread near the Virunga Volcanoes in the prefectures of Ruhengeri and Gisenyi (African Rights 1998). According to one unconfirmed estimate, the area around Rwanda's Parc National des Volcans (PNV) was the scene of ten murders each night during 1997. Anyone working with the RPF was considered a collaborator and became a target of the *interahamwe*, including national parks staff. Murders continued but were less frequent in 1998. It was not until 1999 that peace was restored to most of the country.

Democratic Republic of the Congo

Following the successful offensive in October 1996 to close down the refugee camps, Rwandan forces supported Laurent Kabila as the Congolese leader of a movement that aimed to overthrow President Mobutu. With support from Rwanda and Uganda, Kabila took control of the whole country within a period of about ten months. During this war, Kabila was often accused of being a "puppet" of Rwanda and Uganda. According to some reports, many Hutus were massacred at the beginning of the war when they fled from refugee camps into the forest. The United Nations (UN) established an international investigation team to look into these allegations, but their work was repeatedly obstructed and ultimately they were unable to visit the sites where the massacres allegedly occurred. Interestingly, this harmed Kabila's international reputation rather than that of Rwanda or Uganda, despite the fact that he was considered a puppet of both countries.

Once Kabila was in power, Rwanda and Uganda sought his support in demobilizing the small rebel groups and *interahamwe* that existed along their borders. Rebel groups had created insecurity in western Uganda for several years and the *interhamwe* attacks were leading to many deaths in Rwanda. However, President Mobutu had failed to do anything to help. In 1998, friction started to build between Kabila and his allies, and soon he asked the Rwandans to leave the DRC. This led to renewed civil war. Rwanda and Uganda supported the civil uprising in the eastern DRC, while Kabila enlisted the help of Angola and Zimbabwe to halt the new war.

By late 1999, the two sides had effectively partitioned the DRC in two, with Kabila and his allies controlling the west, and Rwanda and Uganda working with Congolese groups to control the east. Several efforts were made to bring all sides to meet and talk about peace, but they were largely rebuffed. In January 2001, Kabila was assassinated by members of his own guard, and consequently there are renewed hopes that peace may be restored. Nevertheless, insecurity persists on both sides of the front line, including areas that are supposedly controlled by one of the factions.

IMPACT OF THE WARS ON PROTECTED AREAS

Rwanda

Virunga Volcanoes–The Virunga Volcanoes comprise approximately 425 km^2 of forest and open parkland located between 2,600-4,500 m. The volcanoes straddle three countries, and have national park status in each one: Parc National des Volcans (Rwanda), Virunga National Park (DRC), and Mgahinga National Park (Uganda). This region is best known for its mountain gorillas (*Gorilla gorilla beringei*) and the Karisoke Research Center (KRC) that was established by Dian Fossey and whose researchers have studied the gorillas for over 30 years.

In 1980, following research and project design by the WCS, the Mountain Gorilla Project was established with support from Fauna and Flora International, the African Wildlife Foundation, and later the World Wide Fund for Nature (WWF) (Vedder and Weber 1990). This project created a tourism program, which came to be viewed as a model conservation project. It brought in millions of dollars for Rwanda and ORTPN (Rwandan Office for Tourism and National Parks) and indirectly financed the conservation of the other protected areas in Rwanda. Just before the war, ORTPN was supported by the International Gorilla Conservation Programme (IGCP). IGCP had grown out of the Mountain Gorilla Project and KRC and was supported by the Dian Fossey

Gorilla Fund International (DFGF). The KRC and IGCP had expatriate directors, but IGCP's director was involved in improving collaboration and coordination between Uganda, Rwanda, and the DRC.

Following the invasion by the RPF in 1990, it was not possible to enter the forest in the Virunga Volcanoes region between the mountains of Sabinyo and Muhabura in the eastern sector. That year, government forces arrested the Rwandan park ecologist, a Tutsi, and imprisoned him in Ruhengeri. Attacking RPF forces released him in February 1991, and he subsequently fled the country. Both warring factions (RPF and the then government forces) laid anti-personnel mines in the park and along its border. However, park staff continued to patrol the forest in the west. In 1991, the Rwandan army cut a 10 m wide swathe of vegetation (north to south) across the forest to facilitate patrols along this line and prevent any RPF soldiers from moving into the western half of the park. Although they occasionally fired mortars into the forest at the RPF, both government and RPF forces declared they would not harm the park or the gorillas. The park was also severely impacted when refugees from Rwanda settled nearby in the DRC. Refugees entered the park and cleared a large area of forest for firewood (Henquin and Blondel 1996; Werike, Mushenzi, and Bizimana 1998).

At different times between 1990 and 1994, the deterioration of the security situation forced the evacuation of KRC researchers from their field site in the forest. In 1994, the then director of KRC evacuated to the eastern DRC. Working with the DRC representative for IGCP, he helped establish a refugee camp specifically for KRC employees and Rwandan park staff who had crossed the border. Most of the staff from the park and the KRC returned to Rwanda one month later, when the RPF established control over the country and the security situation improved. While in the camps, many park and KRC staff members lost friends or family to cholera. As staff members returned to Rwanda, some were attacked by the *interahamwe*–who opposed anyone's repatriation. One KRC employee required hospital care for a month due to machete wounds. At that time, all the senior staff for PNV had left Rwanda and it was still too insecure for the director of KRC to be near the park.

During periods when expatriate researchers and the senior staff from KRC had evacuated, the junior staff was the only conservation presence in the field. Consequently, the junior staff took on the responsibility of organizing patrols, monitoring the gorilla groups, and sending people to bring salary payments from Kigali by taxi. As calm resumed, new senior staff members were appointed to ORTPN for PNV and KRC. With financial support provided from IGCP, an international team removed all anti-personnel mines, and eventually tourism to the gorillas recommenced.

Although a study that I undertook of the war's impact on ungulates in PNV showed that ungulate numbers had not changed significantly since 1989 (Plumptre and Bizumuremyi 1996; Plumptre et al. 1997), poaching of wildlife in the park had indeed accelerated. As poachers increased sales of bushmeat, its price declined. At the same time, the decline of livestock and poultry production across the country led to higher prices for domestic meat (Plumptre et al. 1997). During 1997 and 1998, the *interahamwe* targeted KRC staff and caused several deaths. Many people from the outlying region fled their communities and lived in the forest during this time, leading to heavy poaching for bushmeat. Surprisingly few gorillas from the habituated groups were killed during the war, although a census of the population is required to see if this is true for the whole population.

Tourism to the Virunga Volcanoes region dropped precipitously because of the civil war (Figure 2). It started to recover after the genocide in 1994, but ceased again in 1997. Renewed attacks by the *interahamwe* from the park in northwestern Rwanda kept it low in 1998. In 1999, it was again possible to visit the gorillas, and tourism resumed.

Nyungwe Forest Reserve–The Nyungwe Forest Reserve in southwest Rwanda covers 970 km^2 of steep mountainous terrain and is one of the largest remaining montane forests in Africa. It contains many species endemic to the Albertine Rift, particularly birds and primates. This forest was less affected by the war in Rwanda. Little fighting took place in the reserve because French forces occupied it during Operation Turquoise. The Projet Conservation de la Foret de Nyungwe (PCFN), supported by WCS, continued to operate during this period, as did the ORTPN staff. In addition to PCFN, there were other projects in the Nyungwe Forest Reserve supported by bilateral aid from France, Switzerland, and the European Union, but the deteriorating security in Rwanda forced their permanent closure.

During 1994, the expatriate directors and senior Rwandan staff members of PCFN evacuated the forest and left Rwanda, leaving the junior staff to manage the project. The warden employed by ORTPN remained but was subsequently murdered. The junior staff reorganized and continued working. For five months, they labored without receiving any payment of salary (Fimbel and Fimbel 1997). As security improved, the directors of PCFN returned for a short visit and appointed a new warden (a Rwandan ecologist who had worked at KRC). Although initially the

FIGURE 2. The number of tourists visiting Rwanda's Parc National des Volcans (PNV) to see gorillas from 1976-1997.

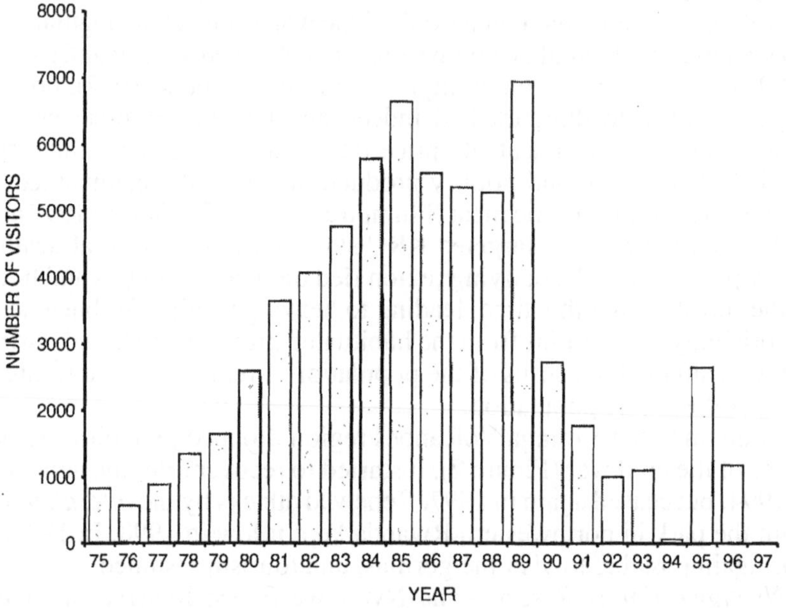

forest was alleged to contain pockets of resistance, particularly near the Burundi border, calm was restored to this region by 1998.

Akagera National Park–Akagera National Park in eastern Rwanda contained 2,800 km^2 of savanna and wetlands. During the initial invasion in 1990, this park was invaded by the RPF but they were quickly driven back to Uganda. Due to the presence of rebel and government military personnel in the park from 1990-93, many animals were killed, primarily for bushmeat (Kanyamibwa 1998). ORTPN was unable to support its staff, and WWF removed all of its personnel early in the war. Following the change of government in 1994, the RPF decided use a large portion of the park for the resettlement of Tutsis who returned with cattle and needed grazing land. Despite international protest, the park boundary was redrawn in 1991, and the government handed over approximately 66% of the park to the returning Tutsis (Figure 1).

Democratic Republic of the Congo

Kahuzi-Biega National Park–What took place in this park is the subject of another paper in this volume, so this discussion will be brief.

Kahuzi-Biega National Park, located in the eastern DRC, contains about 86% of the world's population of eastern lowland gorilla (*Gorilla gorilla graueri*) in an area of about 5,400 km^2. In 1994, when Rwandans fled to the eastern DRC, the warden of Kahuzi-Biega National Park showed great forethought. By negotiating with military leaders and the international aid agencies, he ensured that the refugees were settled only on one side of the park and at some distance from the park's border, unlike the earlier situation around the Virunga Volcanoes. At that time, the German Technical Agency for Cooperation (GTZ) supported the conservation of the park. They maintained support for the park, although at a reduced level. Following the invasion from Rwanda in 1996, many people fled into the park. The security in the region deteriorated and has since remained poor. Few people in the local population support the new regime, and the area is plagued by frequent attacks from *interahamwe* and a group of Congolese known as the Mai-Mai.

The political situation has made it very difficult for the park warden and his ICCN staff. With their national headquarters located in Kinshasa in the western side of the DRC, they live under a different government and consequently receive no pay. Fearing that they could become targets if the government in the west were to regain control of this part of the DRC, the ICCN staff is reluctant to work too closely with the new regime.

At the time of the invasion, the park guards were disarmed by the Congolese faction in this region, the Rassemblement Congolais pour la Democracie (RCD). They maintained some patrols in the forest, but without weapons they were powerless against armed opposition. To address their ineffectiveness, a training program was organized in which the RCD/RPA provided the park guards with some military training. They have since been permitted to carry arms in the park. A census organized by WCS and ICCN in 2000 showed that during the period in which the guards were unarmed, 40% of the gorillas and 95% of the elephants that were present in the park in 1996 had been killed for their ivory and meat.

Okapi Wildlife Reserve–The Okapi Wildlife Reserve was created to protect the okapi (*Okapi johnstoni*), a forest-dwelling relative of the giraffe that is found in 13,800 km^2 of lowland rainforest. Just prior to the war, ICCN managed this reserve with support from Gilman International Conservation and WCS. During the first invasion of 1996, the retreating Congolese army and the invading Ugandan army came through the reserve and looted equipment. Senior Congolese staff members of the projects and ICCN were forced to hide in the forest because they were thought to have money. They depended on the junior staff to hide equipment and support them while in the forest. Following the initial invasion, the senior Congolese staff returned and managed the projects very effectively. They negotiated with each of the incoming regimes and worked to impress upon them the importance of the reserve's conservation.

The disarming of the ICCN staff members hampered their ability to prevent illegal bushmeat hunting in the reserve, particularly of elephants. As the armies passed through during each invasion, the forest was used to supply food to the troops. This has resulted in a decimation of the duiker (*Cephalophus* spp.) populations. Pygmy groups that used to hunt within 20 km of the roads that pass through the reserve now must travel up to 30 km to find any food. The demand for bushmeat has been the primary reason for the hunting in the reserve. Consequently, elephant meat has been selling at higher prices than ivory in this reserve (R. Mwinyihali pers. comm.). Hunting has been far more of a problem in the DRC than in Rwanda. This is partly due to the superior availability of large mammals in the DRC's protected areas, but also because the people living in the DRC have historically hunted many animal species for bushmeat.

SURVEY OF PROJECT STAFF

In 1998, I undertook a survey of the staff of the KRC and PCFN to assess what they had endured during the war and to understand why they had continued working despite great risk to their lives. Each person was interviewed by a trained field assistant (one at each site) using a semi-structured interview process. Local assistants were employed at each site so that interviewees would feel less intimidated in answering certain questions. Furthermore, these assistants were present in the projects during the wars and could check to confirm that interviewees responded truthfully.

The fighting in Rwanda has resulted in the violent deaths of many project staff members and park rangers. Among those tragically killed were ten ORTPN rangers (about 25% of all rangers) in the Virunga Volcanoes; 13 assistants and anti-poaching rangers of KRC (50% of the staff); 44 park rangers (about 50% of the ICCN staff) in the southern section of Virunga National Park (the area that incorporates the Virunga Volcanoes); four ICCN rangers in the Kahuzi-Biega National Park (about 7%) (N. Mushenzi pers comm.); and one ORTPN warden in

Nyungwe Forest (about 2%). Of the KRC staff, 94% had been robbed (compared with 41% of the PCFN staff) at some point during the war, and 88% had lost a member of their family (compared with 43% of the PCFN staff). Few members of the PCFN staff believed that their work had increased any risk to their life (14%), but 69% of the KRC staff believed that their lives were at greater risk. This was because people around PNV were targeted by the *interahamwe* who believed they worked with the RPF government.

When asked what motivated them to continue working despite the risk to their lives, salary was obviously a prominent response (Table 1). The regular payment of the salary was noted as being more important than the total amount of the salary. Regular payments allowed the staff to survive when crops were stolen or not planted. The fact that KRC/PCFN cared for their employees and helped look after them was also an important motivating factor in keeping them working when no salary

TABLE 1. Reasons Given by Project Staff Members When Asked Why They Continued to Work Despite the Risk to Their Lives and What Suggestions They Had for Improving Projects in Times of Insecurity

The percentage of respondents that suggested a reason is given for each project separately. These global responses have been synthesized from more detailed responses that differed slightly from one individual to the next.

	PCFN (n = 60)	KRC (n = 17)
Why you continued working?		
Salary	90.1	94.1
Like the nature of the work	52.5	47.1
Work is important for Rwanda	57.4	47.1
The project looked after my family/me	11.5	35.3
Education possibilities	6.6	11.0
Senior staff members are committed to us	4.9	0.0
What can projects do?		
Help employees avoid risk	83.6	94.1
Collaborate with authorities but be neutral	26.2	52.9
Listen to employees' suggestions	3.3	5.9
Financial help for reconstruction	16.4	11.8
Improve communications–radios	16.4	11.8
Suspend work if too dangerous	16.4	0.0
Educate the local population about project	68.8	0.0

was forthcoming. They believed that this indicated that the senior project staff would eventually return. The possibility of further education was cited as a motivating factor as well. Many respondents also stated that they liked the nature of their work. They felt it was a worthwhile vocation, and considered that it was important for their country.

The latter reason shows that their motivation to continue working was affected by an understanding of the importance of conserving their country's forest. Before the war, conservation education messages were frequently broadcast on the radio and promoted in Rwandan schools. Beginning in 1979, an environmental education program operated around PNV. It included the presentation of talks and films in villages. Consequently, most people in Rwanda were familiar with the importance of the forests for gorillas and watershed conservation (Weber 1987).

When asked what projects could do to help their staff during war, the most frequent response recommended improving the safety of their staff (Table 1). Specific suggestions included helping staff members find a safer place to live, or alternatively, a safer place to work. A few of the PCFN staff members suggested suspending work if these measures were not possible. However, most staff wanted to try to continue working despite the dangers. During the upsurge in violence in Ruhengeri in 1997-98, many members of the KRC staff were housed in there to protect them and their families. At that point, they worked in the park with the military during the day. Respondents also considered it important to collaborate with the authorities in power, but to avoid taking sides. Other suggestions included improving communications with the provision of radios, meeting with the other staff wherever possible, and educating the local communities during peaceful interludes about the importance of the forests for their long-term livelihoods. WCS and IGCP have since provided radio systems to the staff of PNV and the Nyungwe Forest Reserve.

In Rwanda, arguments made in favor of conservation have focused on the economic benefits of increased foreign exchange from tourism. They have rarely cited the people's national pride in their gorillas. However, national pride may be an important factor in gorilla conservation, particularly during times when tourism is not possible and income from foreign exchange drops. Survey results show that the pride the PCFN and KRC staff felt in protecting a part of their national heritage was one of the factors that kept them working during the civil war. Another factor was the global importance of their forests for conservation. This is often the motivation for conserving animals in wealthier countries as well. The national parks agencies in Rwanda (ORTPN), the DRC

(ICCN), and Uganda (Uganda Wildlife Authority–UWA) should all aspire to foster these positive attitudes toward conservation. Education programs can be an important component of efforts to bolster national pride in conservation.

LESSONS LEARNED

Is conservation a luxury to be pursued in areas unaffected by armed conflict? If not, how can we better protect reserves and parks during times of conflict? Given that some regions are more likely to be affected by war and unrest, how can conservationists prepare for it? Considering these questions, what are the lessons that can be drawn from the examples presented here?

Importance of junior staff–When civil war started, most of the senior staff members of these projects became targets and were compelled to flee. Armed groups targeted the senior staff for various reasons: (1) they had worked closely with the previous regime; (2) they were perceived as wealthy or having access to project funds; or (3) they were believed to be supporters of the incoming regime. In some cases, it was possible to support training of senior national staff members during the time that conflict prohibited their return to the field.

During most of the periods of greatest insecurity, junior staff members were left to manage as best as they could. For the most part, they did a very good job despite the fact that before the war few of them had been given major responsibilities (Hart 1997; Hart and Hart 1997; Fimbel and Fimbel 1997). Before the genocide, it was PCFN policy to give leadership roles to junior staff members who showed promise. These junior "crew leaders" were the ones who took on greater responsibilities during the insecure periods (C. Fimbel pers. comm.).

The demonstrated potential of junior staff has changed thinking in WCS about the value of broader training for junior staff. Training should be designed as either courses or on-the-job mentoring that enables certain individuals in the junior staff to support senior staff by managing small projects within the main conservation program. Most conservation projects provide training to senior staff and possibly some basic training to junior staff, but they rarely consider preparing junior staff to assume responsibilities. Likewise, conservation projects rarely consider facilitating graduate-level university training for junior staff.

There are some limitations to the role that junior staff can play in lobbying politicians to protect a reserve, but during times of conflict they can play a vital role in short term efforts. With the support of non-governmental organizations (NGOs) or a government institution, they can often undertake much of the necessary political lobbying while senior staff are absent from the field.

Importance of maintaining a presence–When war started in Rwanda, increased threats to expatriate staff caused many of the bilateral projects to suspend their activities and leave the country. There is now little to show for the millions of dollars that were spent on these projects. In Akagera National Park, external support to ORTPN was withdrawn with no indication of when it might resume. When the new regime was established, no staff remained to advocate its conservation, and consequently it suffered much more than PNV or the Nyungwe Forest Reserve. Gishwati Forest Reserve (Figure 1) had no significant NGO presence before the war. A survey in 2000 showed that the forest, which had been the size of the Virunga Volcanoes in the early 1980s, had been converted completely to farmland (WCS unpublished report). In the DRC, there were few bilateral projects because Mobutu's government had been "blacklisted" from most international aid, and therefore the withdrawal of funding was less of a problem. However, the war created problems for NGOs that tried to raise funds for projects in the DRC.

The staff of KRC and PCFN believed that their conservation organizations would return to support their work. The perceived commitment of KRC and PCFN was an important factor that motivated the staff's continued protection of the forest, even as violence claimed the lives of many of their co-workers. If conservation is to be successful in areas affected by conflict, longer-term vision and commitments (but not necessarily more investment) are required from major donors.

So much international aid today is fickle. Projects are commonly supported for 3-5 years with huge sums of money because it is considered uneconomic to give smaller amounts of money over longer periods. Following those 3-5 years, support is transferred to another areas, and the project is left to struggle on its own. Many international consultants and nationals who take jobs with projects understand this reality. It is a challenge to maintain trained staff on such projects because they know that funding will run out after only a few years. In the case of several projects in the Nyungwe Forest Reserve, staff members all left, probably because they correctly assumed that the projects would not resume after the genocide.

Most projects do not need quick infusions of millions of dollars, but instead would benefit from these amounts spread over decades. Furthermore, given the dangers associated with moving large sums of money to a project during times of conflict, it is difficult to maintain large projects amidst insecurity. However, both in Rwanda and the DRC it has been possible to provide smaller levels of support at regular intervals during the conflicts. With long-term goals and work plans, project staff will believe that continued funding is likely. They will therefore continue working without salaries in the belief that they will be compensated in future. Without long-term plans, the continued support of staff during difficult times is unlikely.

Importance of maintaining neutrality–The local staff of PCFN and KRC cited maintaining neutrality as a high priority for future projects working under similar conditions. This was also an important consideration for staff working with WCS in the Okapi Wildlife Reserve. It is equally important that foreign conservationists maintain neutrality when operating abroad in areas of violent political conflict.

At the same time, it is important that projects and staff collaborate–or at least coordinate–with the authorities in power. From 1997-98, KRC and ORTPN staff members were not perceived as being neutral because they patrolled the forest with military escort as a necessary safety precaution. This perceived lack of neutrality led to the deaths of several employees. Therefore, great care should be taken in making decisions to work with the military. Projects that choose collaboration should consider providing increased protection to their employees as done at KRC.

The right balance between maintaining a neutral position and collaborating with the authorities is difficult to achieve. It is also difficult to judge whether you are succeeding. Outside of the work environment, it is even more difficult to maintain neutrality when staff socialize with authorities or members of a new regime. This is particularly true if they only socialize with individuals from one of the sides present within their community.

Listen to and care for employees–At least 61 conservation employees have been killed while working to protect the reserves discussed in this paper. Little has been done to support the families that have lost the main breadwinner. IGCP has helped the families of slain staff in the Virunga National Park in the DRC, and friends of KRC have done the same for the survivors of KRC employees who have died. The Okapi Wildlife Reserve and KRC have demonstrated leadership on this issue by renting housing in "safer areas" to protect project staff and their families. This has helped strengthen the commitment felt between staff and the project. Another step forward would be the creation of a fund to support the families of conservation employees who are killed as a result of war or poaching activity.

Some project staff interviewed in the study emphasized the importance of listening to the advice of junior staff members. They felt that at times this had not happened. In many cases, junior staff members were in a better situation to judge the security situation, and it was important that senior staff did not push them to do something they considered too risky. As the war drags on, senior staff members often find it difficult to suspend work because there is increasing pressure to show progress to donors. At times it may be necessary to suspend work completely in order to protect the people in whom the project has invested time and money.

Importance of good communication–Anyone from the military will recognize the importance of good communications. The conservation staff of PCFN and KRC readily identified communication as being crucial. Interviews with project staff in 1998 highlighted the importance of radio systems. Existing radio systems were looted during the war and the ensuing periods of insecurity. Consequently it was very difficult for the junior staff to communicate with the capital, 'safe towns', or other project personnel in the forest. It is important to install new radio systems as soon as possible. As previously mentioned, WCS and IGCP have since provided radios for PNV and the Nyungwe Forest Reserve. GTZ has similarly equipped rangers in Kahuzi-Biega National Park. Satellite telephones can also help facilitate communication and are often easier to hide from armed groups and looters. However, the high cost of using a satellite system may permit only occasional communication, rather than regular contact.

Education of the local population–Community education projects exist in many sites around protected areas, but it is often difficult to measure their contribution to conservation. The education program of the Mountain Gorilla Project was successful in raising awareness about mountain gorillas during the 1980s, but this does not necessarily mean that it changed attitudes towards gorillas.

For ten years, armed conflict and insecurity have exacerbated the problem of poaching in the forest of the Virunga Volcanoes. Snaring of antelopes was frequent, particularly during 1997-98. However, despite the fact that people were living temporarily in the forest and some probably faced starvation, few gorillas were killed. Not far away, in the Kahuzi-Biega National Park, hunting pressure has been more intense. It appears that at least 40% of the gorillas in the mountain sector of the Kahuzi-Biega National Park have been killed. Why the difference?

Regional cultural differences may help to explain these contrasting experiences. Unlike the Congolese near Kahuzi-Biega National Park, the Rwandans and Congolese adjacent to the Virunga Volcanoes generally do not eat primate meat. However, it is possible that people living in the forest in the Virunga Volcanoes would have eaten anything to survive.

Another important factor was the widespread educational campaign in Rwanda during the 1980s that emphasized the national and global importance of the gorillas. When I worked in Rwanda in the late 1980s, many Rwandans were demonstrably very proud of the gorillas. Gorilla stickers, bars of soap, posters, postcards, notebooks, and carvings were common, and today there is a gorilla hologram on the Rwandan travel visa.

Among the local people around Kahuzi-Biega National Park, the popular attitude toward gorilla conservation was quite distinct. Many people resented the presence of the park because it denied them access to cultivatable land. Before the war, corruption was widespread and many authorities did not respect the law. The police and the army often extorted bribes and were considered among the most corrupt. When the war led to the removal of government authorities, people decided to take what they had always been denied.

Neither PNV nor Kahuzi-Biega National Park had a significant community conservation program underway, although the GTZ project around Kahuzi-Biega National Park was probably more advanced than around PNV. Therefore, discrepancies in the community benefits derived from each protected area do not provide a likely explanation for their contrasting experiences. I believe that the national pride in gorillas that resulted of Rwanda's education programs was the main factor in the survival of PNV's gorillas.

CONCLUSIONS

Political violence and armed conflict will continue to affect protected areas. Although each conflict is unique and the impacts are difficult to predict, conservation organizations can take steps to better prepare for civil strife and insecurity. This paper has identified specific measures that can help mitigate the impact of armed conflict on conservation projects, wildlife, forests, and local communities.

Surveys of project staff show that the greatest priority for organizations should be to recognize the critical role of junior national staff. Organizations should develop training programs that enable them to assume major responsibilities, exercise initiative, and administer small projects independently. Developing contingency plans for different levels of security risk can also help staff prepare for conflict. With an understanding of how their project hopes to support them, field staff will be more likely to continue working when suddenly faced with a crisis. By maintaining a presence, however small, projects can make an enormous contribution to the long-term survival of a protected area. In the field, staff can make efforts to convince combatants and military leaders to respect protected areas or even actively collaborate in their conservation. However, it is vital that project staff maintain their neutral status while working with such authorities. Communication systems greatly facilitate field operations, and despite the risks of theft, radios should be deployed wherever possible. Finally, there is some indication that education of the local community around a protected area can help discourage poaching and forest destruction. It is during times of armed crises that we are most likely to see whether education programs really have succeeded in creating positive attitudes toward conservation.

REFERENCES

African Rights 1998. Rwanda. The Insurgency in the Northwest. African Rights, London.

Berry, J.A. and Berry, C.P. 1999. Genocide in Rwanda. A Collective Memory. Howard University Press, Washington, DC.

Fimbel, C. and Fimbel, R. 1997. Conservation and Civil Strife: two perspectives from Central Africa. Rwanda: the role of local participation. Conservation Biology 11: 309-310.

Hart, J. 1997. Conservation in Crisis. UNESCO Sources 95:14.

Hart, T. and Hart J. 1997. Conservation and Civil Strife: two perspectives from Central Africa. Zaire: New models for an emerging state. Conservation Biology 11: 308-309.

Henquin, B. and Blondel, N. 1996. Etude par Teledetection sur l'Evolution Recente de la Couverture Boisée du Parc National des Virunga. Unpublished report, Laboratoire d'hydrologie et de Teledetection, Gembloux, Belgium.

Kanyamibwa, S. 1998. Impact of war on conservation: Rwandan environment and wildlife in agony. Biodiversity and Conservation 7: 1399-1406.

Plumptre, A.J. and Bizumuremyi, J.B. 1996. Ungulates and hunting in the Parc National des Volcans, Rwanda. The effects of the Rwandan civil war on ungulate populations and the socioeconomics of poaching. Report to the Wildlife Conservation Society.

Plumptre, A.J., Bizumuremyi, J-B., Uwimana, F. and Ndaruhebeye, J-D. 1997. The effects of the Rwandan civil war on poaching of ungulates in the Parc National des Volcans. Oryx 31: 265-273.

Prunier, G. 1995.The Rwanda Crisis. History of a Genocide. Hurst & Co., London.
Vedder, A. and Weber, A.W. 1990. The Mountain Gorilla Project (Volcanoes National Park). pp. 83-90 in A. Kiss (ed.) Living with Wildlife. Wildlife Resource Management with Local Participation in Africa. World Bank Technical Paper 130. Washington, DC.
Weber, A.W. 1987. Socio-ecological factors in the conservation of afromontane forest reserves. In: Primate Conservation in the Tropical Rain Forest. pp. 205-229 in C. Marsh and R. Mittermeier (eds.). Alan R. Liss, Inc., New York.
Werikhe, S.E.W., Mushenzi, N. and Bizimana, J. 1998. L'impact de la guerre sur les aires protégées dans la région des Grands Lacs. Le cas de la région des volcans Virunga. Cahiers d'Ethologie 18: 175-186.

Building Partnerships in the Face of Political and Armed Crisis

Annette Lanjouw

SUMMARY. The crisis in the Great Lakes region has affected Rwanda and the Democratic Republic of the Congo (DRC) since 1990, but its roots reach back into the colonial and pre-colonial past. Decades of conflict in the region have affected livelihood strategies of local people and caused enormous population displacements, all of which have had impacts on the natural environment and protected area management. Conservationists working to protect and effectively manage natural resources and protected areas have much to learn from the experience of humanitarian and relief organizations working in conflict situations. In addition, relief and development organizations can learn from some of the approaches applied by conservation agencies. In this paper, lessons from the experience of the humanitarian sector are analyzed and their value for conservation organizations working amidst political and armed crisis is examined. This analysis draws upon the experience of the International Gorilla Conservation Programme (IGCP) in the DRC and Rwanda from 1991 to date. Recommendations are made for greater collaboration and programmatic integration between the conservation, relief, and development sectors. *[Article copies available for a fee from The Haworth Document Delivery Service: 1-800-HAWORTH. E-mail address: <getinfo@haworthpressinc.com> Website: <http://www.HaworthPress.com> © 2003 by The Haworth Press, Inc. All rights reserved.]*

Annette Lanjouw is Director, International Gorilla Conservation Programme (IGCP), P.O. Box 48177, Nairobi, Kenya (E-mail: alanjouw@awfke.org).

[Haworth co-indexing entry note]: "Building Partnerships in the Face of Political and Armed Crisis." Lanjouw, Annette. Co-published simultaneously in *Journal of Sustainable Forestry* (Food Product Press, an imprint of The Haworth Press, Inc.) Vol. 16, No. 3/4, 2003, pp. 93-114; and: *War and Tropical Forests: Conservation in Areas of Armed Conflict* (ed: Steven V. Price) Food Products Press, an imprint of The Haworth Press, Inc., 2003, pp. 93-114. Single or multiple copies of this article are available for a fee from The Haworth Document Delivery Service [1-800-HAWORTH, 9:00 a.m. - 5:00 p.m. (EST). E-mail address: getinfo@haworthpressinc.com].

KEYWORDS. Civil war, protected areas, humanitarian, relief, development, conservation, mountain gorillas, Great Lakes region, Virunga volcano massif

INTRODUCTION

The number of wars and civil conflicts, often referred to as complex emergencies, appears to be increasing globally. In 1995, the United Nations (UN) identified 28 complex humanitarian emergencies affecting some 60 million people around the world (Agency for Cooperation and Research in Development 1995). Many of these wars are being fought in border areas. Throughout the world, international borders were commonly drawn along natural divisions, such as mountain ranges, rivers, and lakes. These same features also provided the biological justification for the establishment of many protected areas along national borders. The convergence of these factors has meant that many recent wars and civil conflicts have been fought near, or in, protected areas.

Wars are also increasingly leading to civilian casualties. During World War I, only approximately 5% of casualties were civilians. In World War II, this figure rose to 50%. Civilians now represent 80% of modern war casualties, a large fraction of which is made up of women and children (Ingram 1994). In many wars and armed conflicts, violence against civilians is a deliberate strategy, not an accidental side effect. Frequently, the purpose is to kill or expel civilians of another group (Cairns 1997). These phenomena have been witnessed in many of the past decade's conflicts in Africa. This paper considers how recent complex emergencies in the Great Lakes region of Africa have affected civilian populations, natural resources, and the management of protected areas.

THE GREAT LAKES CRISIS WITHIN THE GLOBAL CONTEXT OF CONFLICT

The Great Lakes region includes the eastern Democratic Republic of the Congo (DRC, formerly Zaire), Rwanda, Burundi, western Uganda, and northern Tanzania. This mountainous and highly fertile area has been inhabited by numerous groups of people for centuries. Before the colonial era, these groups were subdivided into clans, some of which established political control over others. These divisions were not established along ethnic or racial lines, but according to political and economic relationships (Chretien and Triaud 1999). The different ethnic divisions that existed among the different social groups in the area that is now Rwanda, Burundi, and Congo were taken advantage of during the colonial period for political objectives. These divisions deepened after independence, and as a consequence, numerous "ethnic" clashes have occurred throughout the region in the past 50 years (Lanjouw et al. 2001).

Clashes between Hutu and Tutsi ethnic groups in Rwanda during the late 1950s led many Tutsi to flee to Uganda and other neighboring countries. These refugees were never fully integrated, and they remained as a "diaspora" in their new host countries. Repeated clashes and violence against the Tutsi in Rwanda in 1963, 1967, and 1973 resulted in the flight of many more people to neighboring countries. Strains and conflicts between other groups in the DRC also contributed to the tension in the region. The Masisi and southern Kivu regions in the DRC have repeatedly seen clashes between Congolese groups and groups of Rwandan origin ("Banyarwanda"), as well as other ethnic groups.

In Uganda, past problems under the Amin and Obote regimes also led to the movement of many refugees into the DRC and northwards into Sudan. The wars in Sudan, Somalia, and elsewhere in the region have also affected the border areas with Congo and Uganda, increasing the availability of small arms and light weapons (Boutwell and Klare 2000) and the presence of refugees, militias, and rebel groups in all of these countries. Specifically in the region around the Virunga Volcanoes range (Rwanda, Uganda, and DRC), numerous clashes among different groups have led to population displacement across the borders.

These factors all contributed to an attack in October 1990 by the Rwandan Patriotic Front (RPF), from Uganda into Rwanda. The RPF steadily advanced on the capital, Kigali, and in June 1994, President Habyarimana of Rwanda was killed. This triggered a carefully prepared genocide that killed up to a million people in the space of 100 days. The arrival of the RPF in Kigali in July 1994 caused about two million mainly Hutu people to flee into the DRC, Burundi, Tanzania, and Uganda (Joint Evaluation of Emergency Assistance to Rwanda 1996). Also fleeing from Rwanda were the army of the assassinated President Habyarimana and the perpetrators of the genocide–the extremist militia known as the *interahamwe*. The refugees spent more than two years in camps. During that time, the former members of the Rwandan Armed Forces (FAR) and the *interahamwe* regrouped and formed political and military groupings intent on recapturing control of Rwanda (Jongmans 1999). The insurgency that followed greatly disrupted the border regions. It continues today, with incursions by the different militia groups into northwestern Rwanda (African Rights 1998).

At the end of 1996, the dismantling of refugee camps, first in the DRC and then in Tanzania, prompted the forced and rapid repatriation of over two million refugees to Rwanda. This was followed by the deterioration of the security situation inside Rwanda. The country had to grapple with the formidable challenges of resettlement, reintegration, and reconciliation in a post-genocide climate, while attacks continued from rebels based in the DRC (African Rights 1998). In the DRC, the Alliance des Forces Démocratiques pour la Libération du Congo-Zaire (AFDL) began a military operation in mid-1996 that took over the country in May 1997. A new rebel force, the Rassemblement Congolais pour la Démocratie (RCD), launched an attack on Kinshasa the following year. This war continues, with the RCD subdivided into three groups–RCD, RCD-MLC (Mouvement pour la Libération du Congo), and RCD-ML (Mouvement de la Libération)–and supported by troops from neighboring countries. Throughout this period, Uganda has also been affected by movements of Rwandan refugees, insurgencies from the DRC as a result of the war in eastern Kivu, and an escalating cross-border conflict with rebel groups based in the DRC and Sudan (Lanjouw et al. 2001).

The conflicts along the borders between the DRC, Rwanda, and Uganda have not been resolved (Duly 2000). *Interahamwe* militias still roam the forests of the DRC and rebel groups based in the DRC continue to attack Rwanda and Uganda. Clashes between different groups continue to destabilize the DRC. The conflict between the Rwandan and Ugandan-backed rebels in the eastern DRC and President Kabila's forces in the west ensures that political and military objectives continue to be at the forefront of the government's agenda. To date, seven African nations and numerous rebel groups are engaged in the conflict in the DRC (International Crisis Group 2000a).

THE HUMAN CONTEXT OF THE GREAT LAKES CRISIS

During civil and armed crises, deliberate strategies are often implemented to cause famine or disease, or to divert food supplies from certain groups to others. Famine frequently kills more people than violence. The destruction of civil society and the upheaval of populations lead to food shortages. Insecurity can contribute to famine by preventing people from working their fields, traveling to markets, or carrying out other essential subsistence activities. Likewise, the restriction of emergency aid and the deviation of humanitarian support by armed groups can exacerbate hunger (Amartya-Sen and Dreze 1989).

The complex emergency in the Great Lakes region, with its numerous conflicts, has taken an enormous toll on the civil population of the region (International Crisis Group 2000b). While much of the assistance to the region has been focused on the refugees and displaced people, the impact of war on the communities that received the refugees may in fact have been far worse. Many of the host communities in the DRC already lived in extreme poverty or at a subsistence level, and the refugee crisis of 1994-96 severely affected the natural resource base upon which they depended. With their normal means of subsistence disrupted or destroyed, both resident and displaced people became more dependent on local natural resources, including those found in protected areas. The displaced people that settled around the Congolese town of Goma lacked shelter, cooked on open fires, and desperately searched the landscape for food and building materials. Environmental impacts associated with the refugee crisis included heavy deforestation, depletion of fresh water sources, soil erosion, and problems with the disposal of waste and corpses. Consequently, the refugee crisis contributed to food shortages, high prices for firewood, and the spread of numerous diseases (including cholera, dysentery, and venereal diseases).

The host communities faced additional economic hardships when Rwandan refugees took almost every unskilled job in Goma during 1994 (Cairns 1997). With free medical care and food provided to them by humanitarian agencies, refugees were able to accept salaries far below the minimum that was standardly accepted by the local Congolese people. In such situations, differential access to humanitarian assistance can sharpen the existing inequities in host communities. The Congolese population is still struggling to recover from the refugee crisis.

THE IMPACT OF WAR ON THE ENVIRONMENT AND PROTECTED AREAS

The impact of the crisis in Rwanda and the eastern DRC on wildlife, biodiversity, and national parks has been described in numerous reports and articles (Kalpers and Lanjouw 1999; Biswas et al. 1994; Henquin and Blondel 1996). The threats to the environment from the crisis include:

- Destruction or collapse of the economy and loss of opportunity for people to earn a licit livelihood
- Destruction of social structures and institutions, as well as the legal framework, with long-term consequences for the livelihoods of people
- Increased dependence of many different groups (local communities, displaced people, armies, militias, and rebels) on natural resources for food, building materials, firewood, and charcoal
- Presence of armed combat, landmines, and booby traps in forests and protected areas
- Politicization of all work in and around the conflict zone, including park management activities
- General insecurity due to combat, banditry, and the presence of armed groups–making normal activities unsafe and difficult
- Clearing of forest cover by armed factions for military objectives
- Settlement of displaced persons in natural areas/protected areas
- Temporary settlement of rebels/militias in protected areas
- Conditions of lawlessness that allow the staking of illegal land claims by agricultural populations

These threats to the forests and parks led to a number of direct impacts:

- Loss of natural habitat from agricultural encroachment and deforestation for firewood and construction material (in Virunga National Park [PNVi] 105 km^2 of park was affected by deforestation)
- Illegal harvest of natural resources and poaching of wildlife (this included the killing of 18 mountain gorillas from 1995-98)
- Transmission of human diseases to wildlife from people moving through protected areas and the improper disposal of human and medical waste
- Threat to local people and park staff from armed groups inside the protected areas (this can include researchers and conservationists, as well as tourists)
- Increased casualties of wildlife from mines and encounters with armed groups
- Loss of protected area staff to armed conflict (at least 5 park staff in PNVi have been killed by mines and more than 30 killed as a direct consequence of the war since 1995)

- Disarming of park staff by armed groups; leading to increased vulnerability to armed rebels/militias, as well as ineffective protection of the park from illegal activities
- Increased vulnerability of local people (resident populations, displaced people, and park staff) to attacks and raids by armed groups
- Inability of park staff to work effectively due to: a lack of perceived neutrality, disarmament of park rangers, inability to patrol without accompaniment by soldiers (who patrol for military rather than conservation objectives)
- Loss of tourism revenue for the national economy and park management

Many protected areas have been created without the participation of local communities. Their establishment has often involved the eviction and repression of local people, and communities are commonly prohibited from using the natural resources found within the protected areas. Consequently, many protected areas have been the focus of local grievances. Longstanding resentment may constitute another significant threat to protected areas during periods of conflict. Protected areas can be perceived even more negatively when they harbor armed militias or function as a transit corridor for armed groups that raid, ransack, and kill neighboring populations.

The Impact of the Refugee Crisis of 1994-96 on Virunga National Park

Virunga National Park (PNVi)–known initially as Albert National Park–was created in 1925 and is the first national park in Africa. The United Nations Educational, Scientific and Cultural Organization (UNESCO) declared PNVi a World Heritage Site in 1979, and in December 1994 named it a "World Heritage Site in danger." PNVi comprises a unique range of habitats ranging from high-altitude forests and afro-montane habitats to lowland forests, lava flows of various ages, grassland and savannas, wetlands, and even glaciers (Ruwenzori massif). These ecosystems harbor exceptional biological diversity, including a great number of endemic species of plants and animals. PNVi was initially created to protect the critically endangered mountain gorilla (*Gorilla beringei beringei*) (Hilton-Taylor 2000), but it also contains a small population of eastern-lowland gorilla (*G. b. graueri*). The park comprises a total area of about 7,800 km^2 and is subdivided into a number of sectors. The southern sector covers approximately 1,000 km^2 and was the portion of the park most heavily impacted by the refugee crisis of 1994-96 (Kalpers et al. 1999).

In July 1994, following the war in Rwanda, hundreds of thousands of refugees crossed into the DRC and settled around the town of Goma. There they found what they were desperately seeking: water, firewood, and food; all provided by Virunga National Park (PNVi) and its immediate surroundings. On one day alone that month, over 500,000 refugees arrived. Within a few days, another 300,000 had joined them. During July, three camps (Kibumba, Mugunga, and Katale) emerged where the refugees had stopped (Kalpers 2001). To accommodate the continued influx of refugees and to decongest some of the initial refugee locations, two additional camps, Lac Vert and Kahindo, were developed in late 1994 and early 1995. In late 1994, the population of the refugees was estimated at around 750,000. All five camps were established to shelter the refugees from Rwanda and were managed by a number of humanitarian agencies under the overall coordination of the UNHCR.

Deforestation–Loss of forest has been the most visible environmental impact of the refugee crisis. During the more than two years that the refugees remained in the Goma area, large tracts of forest were systematically destroyed, especially in the southern sector of PNVi. The deforestation around the camps was driven by the refugees' need for cooking fuel and construction material, as well as the commercial demand for timber and charcoal in Goma (Languy 1995). This commercial activity was able to flourish due to the state of insecurity in this sector the park.

In less than two years after the arrival of refugees, tree cutting affected 105 km^2 of the park, of which 35 km^2 were completely deforested. The area affected by various degrees of cutting is equivalent to a clear-felled area of 63 km^2 (Henquin and Blondel 1997). Table 1 distinguishes the deforestation estimates associated with each of the refugee camps. Table 2 describes the evolution of deforestation over the 2 years it was evaluated (Henquin and Blondel 1997). Most of the deforestation took place in the more recent lava plains (at 1,500-1,800 m), thus having comparatively little effect on the mature and more biologically diverse forests found on the slopes of the Virunga Volcanoes (at 1,800-3,500 m). However, some mature forest was affected, especially near the Kibumba camp, where there was irreversible damage to the montane *Podocarpus milanjianus* forests.

The forests closest to the Kibumba camp–which held 190,000 refugees–suffered severe and extensive deforestation during the first year. The prominent ecological value of these forests motivated considerable efforts to protect this sector of the park during the second year. Further

TABLE 1. Deforested Areas (km²) Inside Virunga National Park (DRC) Two Years After the Arrival of Rwandan Refugees

Zone	Katale-Kahindo Camps	Kibumba Camp	Mugunga-Lac Vert Camps	TOTAL (5 camps)
Affected area	14	35	56	105
"Equivalent clear-felled area"	6	15	42	63

TABLE 2. Evolution of the Daily Deforestation Rates (Ha/Day) Associated with Refugee Camps Near Virunga National Park (DRC)

Zone	Katale-Kahindo	Kibumba	Mugunga-Lac Vert	Total (5 camps)
1st year of camp	1.4	3.6	5.0	10.0
2nd year of camp	1.1	0.6	6.8	8.4
Average 2 years	1.2	2.0	5.9	9.1

(Deforested area = Equivalent clear-felled area, including deforestation in the buffer zone of Virunga National Park)

damage was practically halted in 1996. In the zone of the Mugunga and Lac Vert camps, where some 200,000 refugees were located, extensive areas were stripped of vegetation. In the other camps, these activities were more effectively controlled and deforestation gradually decreased.

The provision of fuel wood to refugees depleted the tree plantations in the area, thereby expending reserves of wood previously available for use by the local population. The region was already experiencing a shortage of fuel wood, and the arrival of the refugees put an unbearable strain on the region's ability to provide sufficient supplies. Most importantly, this increased pressure on the forests and jeopardized the long-term sustainability of the local population's energy resources.

Commercial bamboo collection and poaching–Aside from the deforestation associated with fuel wood and timber cutting, a variety of other commercial and subsistence activities caused direct environmental impacts in PNVi. Refugees, primarily from the Kibumba camp, were involved in commercial enterprises that required bamboo collected from inside the park. The commercial demand for bamboo caused extensive damage to the limited bamboo zone inside the park, where it is an important seasonal food source for a number of different species of wildlife. Refugees were also heavily involved in poaching, primarily for food. They used traditional snares as well as firearms smuggled into the camps. Poaching quickly became a commercial activity and a significant threat to wildlife populations. The movement of tens of thousands of Rwandan refugees through the forest also caused a great deal of damage and disturbance to the forest. The damage was especially intense when the refugees brought along their herds of livestock.

Disease transmission and health threats to wildlife–The regular movement of insurgents through the park and the presence of refugees greatly intensified the risk of disease transmission to wildlife. Many of the wildlife species–especially the primates–are susceptible to human diseases. The diseases that can be transmitted from humans to gorillas include a number of respiratory diseases (e.g., measles, tuberculosis, pneumonia, and "flu") as well as diseases contracted via the fecal-oral route (e.g., shigellosis, hepatitis, herpes, scabies, intestinal worms, and polio) (Homsy 1999). The periodic presence and passage of thousands of cows, goats, and sheep also posed a health threat for the wild ungulate populations (*Syncerus caffer*, *Cephalophus* spp., and *Tragelaphus* spp.) in the southern sector of PNVi.

In addition, some of the organizations working in the medical sector disposed infectious materials inside the national park. This problem was particularly severe during the first year of refugee operations. These materials included used syringes, human waste, bloodstained materials, and other medical refuse. It is impossible to assess the actual impacts of the unsanitary disposal of medical waste, the presence of livestock, or the movements of refugees and insurgents had on the overall health status of mountain gorillas. Total access to the gorilla habitat is not yet guaranteed, and therefore these impacts can only be measured over time and as "potential" threats.

The Threat of Resettlement in Rwanda's Volcano National Park, 1996-2000

After the Rwandan Patriotic Front stopped the genocide and gained power in Rwanda in August 1994, it was estimated that approximately 50% of the country's population was displaced or settled in only a temporary fashion. In early 1996, a portion of Rwanda's Volcano National Park (PNV), was considered as a site for the resettlement of refugees and displaced people. A commission of the Ministry of Rehabilitation and Social Integration (MINIREISO) officially listed PNV as one of several proposed places for the reintegration of refugees and displaced people. The Arusha Accords, signed in 1993, had excluded PNV from the list of potential zones for the resettlement of old caseload refugees and other people without land. However, pressure to open the park to settlement continued. Since 1995, the population living adjacent to the PNV has made numerous attempts to occupy and settle on lands within the park. The authorities maintained a firm position with respect to protecting the park boundaries.

However, the park boundaries have not always been so steadfastly defended, and over the years, much parkland has been lost. PNV lost approximately 55% of its area between 1958 and 1979; primarily to satisfy the local population's need for land. The ecological value of the park's forests has suffered. The park has lost most of the lower levels of forest vegetation, which form a critical resource for a number of wildlife species, including the mountain gorilla. Scientific data show that poaching (primarily in the DRC) and the conversion of a large portion of Rwanda's PNV into agricultural land during the 1960s led to a 60% decrease in the gorilla population for the Virunga Volcano range (Weber and Vedder 1983). Reductions to PNV have limited the park to those portions that are on steep slopes of more than 15-20%. Any agricultural exploitation of this forest would inexorably lead to severe soil erosion.

Resettlement of displaced people has had deleterious effects on other protected lands in the region, including Rwanda's Akagera National Park (PNA) and its associated Mutara Hunting Reserve. Increased human pressures resulted in the loss of almost two thirds of these protected areas in 1998. Akagera National Park, with its extensive wetlands and humid zones, has been reduced to a small area. It is still threatened by resettlement and urbanization, despite efforts by the international community–particularly the German Technical Agency for Cooperation (GTZ) and the European Union (EU)–to prevent its destruction (Gombe 1995).

In early February 2000, the International Gorilla Conservation Programme (IGCP) learned of a plan to resettle war refugees within Volcano National Park (PNV). The Rwandan government had decided to move approximately 500 of the 8,000 families from a camp of internally displaced persons (IDPs) in the Forest Reserve of Gishwati to PNV. Approximately 627 ha of PNV (approximately 5% of the park) were to be degazetted and given to the families. After numerous missions by park staff and the head of the Ministry of Lands, Resettlement, and Environment Protection (MINITERE), the problem was brought to the attention of the president and vice president of Rwanda. The government was presented with concerns regarding the long-term impacts of the planned resettlement on the livelihood of the local population. After considering the concerns for the potential environmental and socioeconomic consequences, the Rwandan government clearly indicated that the park should remain intact, and that resettlement should take place elsewhere. Although the park is not yet out of danger, the government's desire to protect the park is clear.

PNV has a significant ecological function in Rwanda and the broader region, particularly in terms of protecting water catchment and maintaining soil stability. It also plays an important role in the economy. In Rwanda, tourism is one of the main economic alternatives to agriculture, and the prime tourist attraction is the mountain gorilla. This species also brings a great deal of international attention to the country, and its conservation is therefore a credit to Rwanda in the eyes of the world. In 1989, the gorillas drew more than 7,000 visitors to the country, and thus provided the Rwandan Office for Tourism and National Parks (ORTPN) with more than US$1 million in that year alone. Tourism grew slowly but steadily during the post-genocide period from 1995-97, with associated revenues approaching pre-war levels. Insecurity forced the park to close in 1997, but it was reopened in 1999. Gorilla tourism continues to hold considerable potential for the development of both the national and regional economies.

THE ROLE OF NON-GOVERNMENTAL ORGANIZATIONS WORKING IN AREAS OF ARMED CONFLICT

The role of non-governmental organizations (NGOs) has been much analyzed over the past 20 years. They are responsible for delivering increasing amounts of development and humanitarian/relief assistance. Their ability to deliver this assistance and to function in a politically impartial manner has been debated in light of accusations that they have often had a negative impact on some communities. The role and approach of NGOs in conflict situations has been re-evaluated due to a number of problems: (1) competition between NGOs over "turf" and resources; (2) a lack of coordination between organizations; and (3) allowing emergency assistance to be deviated by warring factions, thus enabling militias to rearm, regroup, and reinforce themselves while supported by "humanitarian assistance."

In just four weeks during August-September 1994, 180 humanitarian and relief NGOs flooded into Rwanda. Over the next two years, 250 NGOs operated in Goma (DRC). These two cases form perhaps the most celebrated examples of the NGO "circus" and are frequently cited to demonstrate the need for coordination, collaboration, and strategic alliances. These criticisms most directly apply to the experience of humanitarian organizations, including NGOs, as well as the United Nations and other multilateral organizations. However, these lessons can also be used to re-evaluate and improve operations within environmental and development sectors.

LESSONS LEARNED FROM HUMANITARIAN AND DEVELOPMENT ORGANIZATIONS OPERATING IN CONFLICT ZONES

A review of literature documenting experiences of NGOs and bilateral and multi-lateral organizations working in complex emergencies and conflict zones has identified a number of important lessons for the conservation, development, humanitarian, and relief sectors (Agency for Cooperation and Research in Development 1995; Bennett and Kayetisi-Blewitt 1996; International Crisis Group 1996). These lessons should help guide the policies and strategies of conservation organizations operating during times of crisis:

- The distinctions between relief, rehabilitation, and development are unclear and often inappropriate for local conditions and communities
- An integrated approach that forges links between humanitarian, developmental, environmental, political, and military sectors is necessary. Such an approach includes relationships between local and external actors (state agencies, outside organizations, and civil society institutions)
- A sustainable strategy for assistance should include investments in local people and institutions
- Strategies for building peace should bring together the interests of warring parties
- Organizational and programmatic flexibility allows groups to respond to evolving situations
- External or technical agencies must be perceived as neutral parties
- Organizations (NGOs as well as governmental agencies and state institutions) must coordinate their efforts

For all of these points, the experience of International Gorilla Conservation Programme (IGCP) has shown that by building partnerships a program can strengthen its ability to remain operational and achieve strategic objectives in times of conflict. Since 1991, IGCP–a coalition of the African Wildlife Foundation, Fauna and Flora International and World Wide Fund for Nature–has been active in Rwanda, the DRC, and Uganda (Figure 1). The program works with the protected area authorities of these countries on four different levels: (1) strengthening the capacity of protected area authorities to manage afro-montane forests; (2) improving collaboration between Rwanda, the DRC, and Uganda on the conservation of their shared ecosystem; (3) increasing support for conservation among various interest groups; and (4) promoting improvements to and respect for policy and legislation related to the conservation of the region's afro-montane forests.

During the Great Lakes crisis, IGCP provided emergency technical and financial support to the region's protected area authorities. The objectives were to help them cope with the enormous challenges that suddenly faced them and to establish mechanisms for mitigating and recovery from impacts of the crises. From 1994-98, IGCP implemented a program for the rehabilitation of Virunga National Park (PNVi) in the DRC and the Volcano National Park (PNV) in Rwanda. Funding support came from the United Nations High Commissioner for Refugees (UNHCR) and other donors.

The distinctions between relief, rehabilitation, and development are not clear–The many organizations that operate during and after times of

FIGURE 1. Central Africa: Democratic Republic of the Congo, Rwanda, and Uganda.

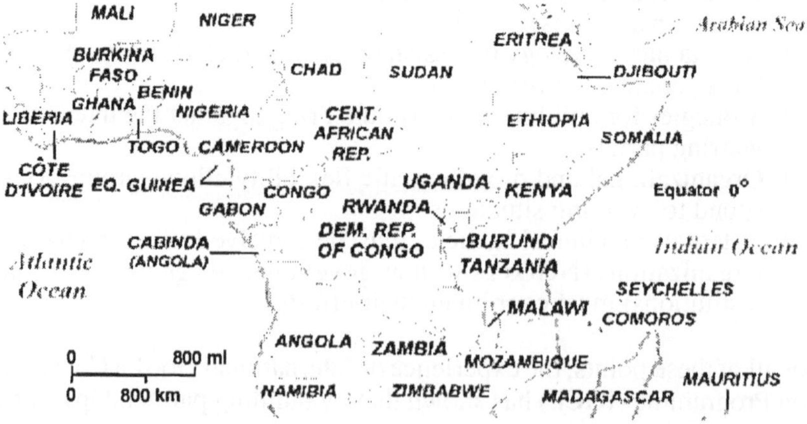

crisis are typically characterized as relief, rehabilitation, or development. However, these services are delivered along a continuum, and the distinctions between the different phases are often unclear and artificial. The distinctions can also be inappropriate for communities whose strategies for survival face long-term crises. A concentration on short-term objectives (typically relief and rehabilitation) can exacerbate development problems over the long term. It is necessary to provide emergency support in a manner that enables communities to achieve long-term development objectives. For example, it is important to avoid enduring problems like those caused by the provision of food and firewood to the refugees in the camps in the DRC from 1994-96. Hastily conceived solutions for the resettlement of displaced people have led to environmental degradation and the loss of long-term agricultural potential for many communities.

From 1996-98, IGCP worked with the UNHCR to rehabilitate the southern sector of PNVi. The rehabilitation program had a strong development component. This included the design of an ecotourism strategy for the park, and more importantly, the initiation of a ranger-based monitoring program that is still used today. Here again, there was a continuum; from the provision of emergency assistance to the strategic strengthening of protected area management and its capacity to cope with future crises.

The park required rehabilitation partly due to the failure of relief organizations to evaluate adequately the long-term environmental impact of the refugees or the importance of the environment to local people. Assistance provided during the refugee crisis was focused exclusively on the provision of relief (Languy 1995). The cost of measures to prevent environmental impacts would have been less than that of the attempted rehabilitation of degraded areas. Moreover, much of the environmental damage was permanent. One positive result of the UNHCR's experiences in the eastern DRC was its design of environmental guidelines for refugee operations (UNHCR 1996). These guidelines emphasize the value of investing in environmental protection, given that restoration of extensively damaged areas is costly and often impossible.

The importance of an integrated approach–During times of crisis, efforts to provide support must link emergency, development, and environmental objectives. Humanitarian assistance must integrate concerns for sustainable development and environmental protection. Emergency assistance should take into consideration the longer-term needs of people and the impact that relief efforts will have on host communities and their environment. During times of conflict, conservation organizations must also situate their objectives in the larger context of human needs and political and military interests. A holistic approach that addresses a whole range of issues is necessary. The focus on conservation objectives in conflict situations is frequently criticized for ignoring human needs and "putting wildlife before people." Although these criticisms are shortsighted, it is important to consider the dependence of local populations on natural resources and link human development and environmental objectives.

The arguments made against the previously mentioned plan to resettle refugees in PNV highlighted the human development aspects of protected area conservation. These include the ecological benefits derived by local farmers and the fundamental role of mountain gorilla tourism in the local, regional, and national economies. As discussed earlier, it was because of these arguments that the government revoked its earlier decision to settle people in the park and reconfirmed its commitment to maintaining the integrity of the protected area.

A sustainable strategy for assistance should include investments in local people and institutions–For many people, economic and environmental crises are far too common. The humanitarian assistance they received from relief organizations may become another finite and perhaps ephemeral resource in an ongoing struggle for survival. The challenges faced by Congolese people during the Mobutu era vary only in degree to those faced in the conflict today. The emergency support provided to populations during recent periods of crisis has not improved people's livelihood strategies or the likelihood that these populations will emerge more capable of surviving future crises. Assistance should support local institutions, ensure local participation, and strengthen the communities' ability to cope with emergencies.

Certainly one of the most critical impacts of war and insecurity observed in the Great Lakes region was the institutional collapse of the protected area authorities. On the whole, they were unable to cope with repeated crisis situations. This was partly due to the weakness of institutions before the crisis. Indeed, protected area authorities will be more likely to endure emergencies if they are strong at the onset of the crisis. An authority's ability to cope becomes even more critical when its external partners must leave the area for security reasons, and it is left to fend for itself.

Experience shown that by investing in the development of organizational capacity and training, programs generally will have greater long-term impact during times of conflict (Agency for Cooperation and Research in Development 1995). Conservation programs should therefore focus on capacity building, both during conflict and in peacetime. Far too often, conservation projects are designed in such a way that they substitute themselves for the official institutions in charge of protected areas. This is typical of short-term projects that concentrate on immediate achievements rather than longer-term capacity.

Building peace by finding and emphasizing the common interests of warring parties–Organizations working on conflict resolution and reconciliation have recognized the importance of identifying common interests between warring parties (Bennett et al. 1996). Conservation can provide an important focus for collaborative processes that build trust and engender peace. However, the potential of conservation to contribute to peace has rarely been thoroughly explored. In areas of conflict where environmental protection is so closely linked with human livelihoods and poverty alleviation, it is critical that this potential be developed.

The Virunga Volcanoes region–which has suffered repeated conflicts during the past decade–provides a promising case. There, three countries (DRC, Rwanda, and Uganda) share the habitat of the mountain gorilla. A regional approach to conservation is therefore critical (Lanjouw et al. 2001). Recognizing that they must work together to ensure that the mountain gorillas are protected in the wild, park authorities of the DRC, Rwanda, and Uganda are working towards building a framework for regional collaboration (Kalpers and Lanjouw 1997). This includes advancing plans for a transfrontier protected area, or "peace park." In the context of the regional crisis, the process of pursuing this goal is as important as its ultimate achievement. By building a framework and process for regional collaboration, conservationists and the development and relief sectors are contributing to the fragile peace process. These efforts reemphasize the observation that peace and reconciliation are not goals, but rather processes (ECCP, IFOR, and CISWF 1999).

The need for organizational and programmatic flexibility–During times of conflict and insecurity, when many planned activities become impossible to implement, programmatic flexibility is extremely important. It can enable organizations to address new challenges where they arise and as they become priorities. To gain greater programmatic flexibility, organizations must step away from the application of a rigid blueprint strategy for their operations. They should instead develop adaptive programs that respond to priority needs and expectations of partners, stakeholders, and beneficiaries, as well as their own interests. Narrowly focused programs often cannot effectively respond to crisis.

During the periods of greatest insecurity, IGCP has worked with the ICCN (in the DRC) and the ORTPN (in Rwanda) to improve training and institutional capacity. The goal has been to strengthen the resilience of projects and programs so that conservation capacity can be maintained during conflict. These priorities emerged as the crisis inflicted heavy losses on protected area personnel and exposed a lack of institutional capacity in the region's protect area authorities. It is also worthwhile to remember that investment in infrastructure or equipment can be destroyed or plundered in a very short time.

In conflict-affected areas, programs can become irrelevant or even inappropriate from one moment to the next. By emphasizing self-reliance and joint planning with the beneficiary and partner organizations, conservation programs can quickly adapt to changing needs and respond to new priorities. Again, conservation organizations can learn from the experience of the relief sector. The classic approach of most conservation programs is to develop strategies that do not include significant contingency plans or consider alternative scenarios. This is characteristic of the "development mentality." Such programmatic approaches are based on the assumption that the situation observed at the design stage of the project will remain stable for the duration of the project. The conservation sector generally lacks the "relief approach," which relies more on operating guidelines, contingency analysis, and frequent tactical changes.

The importance of neutrality or impartiality–Government institutions and NGOs working on the ground in protected areas can never assume that they are perceived as politically neutral or impartial. During periods of armed conflict, the collaboration of conservation organizations with certain communities near protected areas or government institutions (national ministries and park authorities) can be perceived as evidence of impartiality. This can prove dangerous for conservationists in the field. Moreover, NGOs and foreigners are often used as pawns to further the interests of one side or another. For conservationists to operate effectively in the DRC, it is important for all the relevant authorities to understand their goals and objectives and work collaboratively with them.

For the ICCN staff on the ground, the situation is even more difficult. The DRC is effectively divided into sections, each under the control of a different government or rebel authority. The staff of parks located inside rebel-held parts of the country is still answerable to its headquarters located in government-controlled areas. However, any verbal or written report given either to the rebel-government or the central government can be perceived as a demonstration of support for that side. Park staff that travel between the headquarters in government controlled territory to field sites under rebel control may be considered to be untrustworthy and risk imprisonment. To operate under a purely technical mandate in these circumstances is extremely difficult, and the personal risk taken in order to continue operating is enormous. Efforts are being made to contact the Environmental Law Commission in Bonn and other legal agencies to investigate the potential for legally defining a purely neutral mandate and applying "neutral status" to people working for protected area authorities. However, for the people on the ground, the utility of such a status would still depend on the willingness of warring parties to recognize it.

The need for coordinated approaches and collaboration between organizations–NGOs working in the conservation, development, and humanitarian/relief sectors have many complementary activities, but their lack strategic alliances and coordination frequently limits their potential effectiveness. This is due in part to competition between groups for funding, geographic "turf," and programmatic niches. Although these concerns are legitimate, the perceived threats posed by cross-sector coordination are not necessarily real.

However, there are examples of effective coordination and collaboration between organizations and across sectors. In 1999, the ICCN and eight international conservation and development organizations[1] worked together to develop a program for delivering emergency assistance to "World Heritage Sites in Danger in Eastern DRC." The program involves joint planning, harmonized management approaches, and the implementation of joint activities between all five sites. UNESCO will manage the program, with funds provided by United Nations Fund (UNF) as well as financial contributions from each of the eight international organizations involved in the implementation of the program.

Links between conservation organizations and relief organizations, such as the collaborative efforts made by IGCP and the UNHCR to rehabilitate PNVi, are necessary in order to ensure that both long and short-term objectives are accomplished. This collaboration has continued over the years, and in 1997 IGCP was invited by the UNHCR to assist in the development of the UNHCR Environmental Guidelines discussed earlier (UNHCR 1996, 1998). Two years later, IGCP provided training for UNHCR technical staff in Africa on environmental management during refugee operations. This training emphasized the importance of associating local communities with relief activities. These examples demonstrate the potential for bridging the divide between humanitarian/relief objectives and conservation objectives.

It is important to emphasize, however, that coordination is only effective when each party is willing to adapt their programs to a common approach. Accordingly, coordination between the eight organizations working under the UNESCO/UNF project in the DRC has required considerable adaptations to the operations on the ground. On the other hand, the six environmental coordinators that were based in Goma during the refugee crisis of 1994-96 had few–if any–operations on the ground and showed little willingness to fall under the control of any one organization. This did not bring about the intended result: more effective environmental programming.

Continued effective coordination between different organizations working towards conservation objectives can only strengthen the impact of their programs. Coordinated efforts must be focused on practical objectives and results. They should identify the needs on the ground and implement specific cooperative activities to meet those needs. It is important that the overall approach is holistic and responsive to the variety of challenges faced by natural areas and local communities.

While funding for emergency and humanitarian assistance is increasing worldwide, funding for international development is declining. There is also a general decline in public donations to northern NGOs. Given these trends, it is critical for conservation organizations to coordinate their efforts, avoid territoriality, and minimize competition for funds. In addition, the politicization of conservation issues has had an enormously negative impact on the credibility of conservation organizations and their message. Organizations must work together and avoid manipulating information to further personal or individual objectives. Their focus must remain on achieving conservation objectives for the benefit of wildlife, culture, and people over the long term.

NOTE

1. Partners include: IGCP (African Wildlife Foundation, Fauna and Flora International, World Wide Fund for Nature), Wildlife Conservation Society, Gilman International Conservation, International Rhino Foundation, World Wide Fund for Nature, Dian Fossey Gorilla Fund, Zoological Society of Milwaukee, German Technical Cooperation (GTZ) and the Institut Congolais pour la Conservation de la Nature.

REFERENCES

African Rights. 1998. Rwanda: The insurgency in the Northwest. African Rights, London.

Agency for Cooperation and Research in Development. 1995. Development in Conflict: The Experience of ACORD in Uganda, Sudan, Mali and Angola. Oxford University Press, Oxford.

Amartya-Sen and J. Dreze. 1989. Hunger and Public Action. Oxford University Press, Oxford.

Bennett, J. and M. Kayetisi-Blewitt. 1996. Beyond "working in conflict": Understanding conflict and building peace. Relief and Rehabilitation Network Paper, 18. Overseas Development Institute, London.

Biswas, A.K., H.C. Tortajada-Quiroz, V. Lutete, and G. Lemba. 1994. Environmental impact of the Rwandese refugee presence in north and south Kivu (Zaïre). United Nations Development Programme, New York.

Boutwell, J. and Klare, M.T. 2000. Special Report: Waging a new kind of war; A scourge of small arms. Scientific American (June): 46-65.

Cairns, E. 1997. A safer future: Reducing the human cost of war. Oxfam Publications, London.

Chretien, J.-P. and J.L. Triaud. 1999. Histoire d'Afrique: Les enjeux de memoire. Editions Karthala, Paris.

Duly, G. 2000. Creating a violence-free society: the case of Rwanda. Journal of Humanitarian Assistance (January). Retrieved from the World Wide Web, September, 2001: http://www.jha.ac/greatlakes/b002.htm

ECCP, IFOR, and CISWF (European Centre for Conflict Prevention, Implementation Force, and Coexistence Initiative of the State of the World Forum). 1999. People building peace: 35 inspiring stories from around the world. European Centre for Conflict Prevention, Utrecht.

Gombe, A. 1995. Mission Report to Rwanda from 5 February to 13 March 1995. (March) UNEP, Nairobi.

Hilton-Taylor, C. (compiler). 2000. 2000 IUCN Red List of Threatened Species. IUCN, Gland, Switzerland.

Henquin, B. and N. Blondel. 1996. Etude par télédétection sur l'évolution récente de la couverture boisée du Parc National des Virunga.

Henquin, B. and N. Blondel. 1997. Etude par télédétection sur l'évolution récente de la couverture boisée du Parc National des Virunga, deuxième partie (période 1995-1996).

Homsy, J. 1999. Ape tourism and human diseases: how close should we get? A critical review of rules and regulations governing park management and tourism for the wild mountain gorilla, *Gorilla gorilla beringei*. Consultancy for the International Gorilla Conservation Programme.

Ingram, J. 1994. The International Response to Humanitarian Emergencies. In K. Clements and R. Ward (eds.), Building International Community: Cooperating for Peace Case Studies. Allen & Unwin in association with the Peace Research Centre, Canberra.

International Crisis Group. 1996. Towards a crisis prevention plan for Central Africa. International Crisis Group, Nairobi, Brussels.

International Crisis Group. 2000a. Scramble for the Congo: Anatomy of an Ugly War. IGC Africa Report, No. 26. International Crisis Group, Nairobi, Brussels.

International Crisis Group. 2000b. Uganda and Rwanda: friends or enemies? IGC Central Africa Report, No. 14. International Crisis Group. Nairobi, Brussels.

Joint Evaluation of Emergency Assistance to Rwanda. 1996. The International Response to Conflict and Genocide: Lessons from the Rwanda Experience. Steering Committee of the Joint Evaluation of Emergency Assistance to Rwanda, Copenhagen.

Jongmans, B. 1999. Rwanda. In Anonymous, Searching for peace in Africa. An overview of conflict prevention and management activities. European Platform for Conflict Prevention and Transformation in cooperation with the African Centre for the Constructive Resolution of Disputes, Utrect.

Kalpers, J. 2001. Impacts of Ten Years of Armed Conflict in the Virungas Volcanoes Range. Biodiversity Support Program, Washington DC.

Kalpers, J. and A. Lanjouw. 1997. Potential for the Creation of a Peace Park in the Virunga Volcano Region. PARKS: The International Journal for Protected Area Managers, 7(3):25-35.

Kalpers, J. and A. Lanjouw. 1999. Protection of ecologically sensitive areas and community mobilisation. Environmental Management Training Workshop for UNHCR. UNHCR.

Languy, M. 1995. Problèmes environnementaux liés à la présence des réfugiés rwandais. Identification des interventions réalisées. Coordination entre les organismes et propositions d'interventions complémentaires. European Community, Brussels.

Lanjouw, A., A. Kayitare, H. Rainer, E. Rutagarama, M. Sivha, S. Asuma, and J. Kalpers. 2001. Beyond Boundaries: Transboundary Natural Resource Management for Mountain Gorillas in the Virunga-Bwindi Region. Biodiversity Support Program, Washington, DC.

UNHCR (United Nations High Commissioner for Refugees). 1996. Environmental Guidelines. Engineering and Environmental Services Section UNHCR, Geneva.

UNHCR. 1998. Refugee Operations and Environmental Management: Key principles for Decision-Making. Engineering and Environmental Services Section, UNHCR, Geneva.

Weber, W. and A. Vedder. 1983. Population dynamics of the Virunga Gorillas: 1959-1978. Biological Conservation, no. 26:341-366.

Bushmeat Poaching and the Conservation Crisis in Kahuzi-Biega National Park, Democratic Republic of the Congo

Juichi Yamagiwa

SUMMARY. The gorilla and elephant populations of Kahuzi-Biega National Park in the Democratic Republic of the Congo (DRC) have recently suffered from intensive commercial and subsistence poaching. In just four years, the highland sector of the park lost more than 95% of its elephant population and about 50% of its gorilla population. This tragedy was precipitated by the recent political and economic crises in the region. Many factors contributed to increased hunting of wildlife for food known as "bushmeat." Among the principal factors were starvation and economic desperation in local communities, the spread of small arms,

Juichi Yamagiwa is Associate Professor, Laboratory of Human Evolution Studies, Faculty of Sciences, Kyoto University, Sakyo, Kyoto, 606-8502 Japan.

The author thanks S. Price and the other students of the Yale School of Forestry and Environmental Studies for their extraordinary organizational efforts. The author also thanks Dr. S. Bashwira, Dr. B. Baluku, M.O. Mankoto, L. Mushenzi, K. Basabose, and Mbake Sivha for their administrative help and hospitality. The author is also greatly indebted to the guides, guards, and field assistants at Kahuzi-Biega National Park and the members of POPOF for their technical help and hospitality throughout the fieldwork.

This study was financed by the Monbusho (Ministry of Education, Science, Sports and Culture, Japan) International Scientific Research Program (No. 08041146) in cooperation with CRSN (Centre de Recherches en Sciences Naturelles) and ICCN (Institut Congolais pour Conservation de la Nature).

[Haworth co-indexing entry note]: "Bushmeat Poaching and the Conservation Crisis in Kahuzi-Biega National Park, Democratic Republic of the Congo." Yamagiwa, Juichi. Co-published simultaneously in *Journal of Sustainable Forestry* (Food Product Press, an imprint of The Haworth Press, Inc.) Vol. 16, No. 3/4, 2003, pp. 115-135; and: *War and Tropical Forests: Conservation in Areas of Armed Conflict* (ed: Steven V. Price) Food Products Press, an imprint of The Haworth Press, Inc., 2003, pp. 115-135. Single or multiple copies of this article are available for a fee from The Haworth Document Delivery Service [1-800-HAWORTH, 9:00 a.m. - 5:00 p.m. (EST). E-mail address: getinfo@haworthpressinc.com].

and the collapse of park protection during the civil wars. The incentive to hunt gorillas in the park may have developed gradually with the collapse of the Mobutu regime, the influx of refugees, and the two subsequent civil wars. Local resentments felt toward the park and its authorities may have contributed to the willingness of local people to engage in illegal exploitation of wildlife resources. By promoting ecotourism and improving the infrastructure of the park and surrounding villages, the Institut Congolais pour la Conservation de la Nature (ICCN) and the German Technical Agency for Cooperation (GTZ) have made great efforts to improve local conditions. Local people, including park guards and guides, established a non-governmental organization that played an important role in the spread of conservation knowledge and the reduction of conflict among local people. Since the park began employing former poachers in 1999 and resumed conservation activities, gorilla poaching has decreased significantly. The broader civil conflict continues and sporadic poaching is still common. This paper considers and recommends optimal and urgent conservation measures to improve the protection of the park and its gorillas.

KEYWORDS. Democratic Republic of the Congo, large-scale hunting, bushmeat, eastern lowland gorilla, elephant, refugee crisis, conservation, Kahuzi-Biega National Park

INTRODUCTION

History of Kahuzi-Biega National Park

Kahuzi-Biega National Park, in the Democratic Republic of the Congo (DRC)–former Zaire–is well known throughout the world as a center for gorilla tourism where visitors observe wild eastern lowland gorillas (*Gorilla gorilla graueri*) at close range in their natural habitats. Gorillas are one of our closest living relatives and are the largest living primates. Their present distribution is limited to the tropical forests in Equatorial Africa, and the International Union for Conservation of Nature (IUCN) regards them as an endangered species (Hilton-Taylor 2000). Although gorilla taxonomy has been in flux in recent years, *graueri* is a well-recognized subspecies found only in the eastern DRC (Corbet 1967; Groves 1967; Ruvolo et al. 1994), and Kahuzi-Biega National Park is home to the largest remaining population of this subspecies.

The highland sector (600 km^2) was gazetted as a national park in 1970, mainly for the protection of gorillas (Mankoto 1988), but this area also contained important populations of elephant, several monkey and antelope species, and other elements of Africa's montane rainforest fauna and flora. The park's territory, which initially included only montane forest at altitudes of 1,800-3,308 m, was expanded into areas of lowland tropical forest in 1975. It now covers 6,000 km^2 at altitudes from 600-3,308 m, and a narrow corridor links the highland and lowland sectors (Figure 1). Five years later the park was inscribed on the United Nation's list of World Heritage sites (Table 1). In the early 1970s, the late Adrien Deschryver established a gorilla tourism program, and in 1985 the Institut Zairois pour Conservation de la Nature (IZCN) launched a conservation project–in cooperation with the German Technical Cooperation Agency (GTZ)–to train guides and regulate the gorilla viewing program (von Richter 1991).

Status of the Park's Gorilla Population

The first population census on eastern lowland gorillas was conducted in the highland sector of the park in 1978. It recorded 223 gorillas (fourteen groups and five solitary males) (Murnyak 1981). The next census, undertaken in 1990, found 258 gorillas (twenty-five groups and nine solitary males) and suggested that the population of gorillas had been relatively stable or increasing (Yamagiwa et al. 1993). In 1996, the third census found 245 gorillas, including twenty-five groups and two solitary males (Vedder 1996; Inogwabini et al., in prep.). A comparison of census data is shown in Table 2.

According to Hall et al. (1998a), who conducted a population census in 1994 and 1995, the total eastern lowland gorilla population was estimated to be approximately 17,000 individuals. It is possible that most of the population (86%) inhabited the lowland sector of Kahuzi-Biega National Park (Hall et al. 1998b). Between 1,350 and 3,600 elephants were also estimated to inhabit the lowland sector (Hart and Hall 1996). However, elephant poaching was a significant problem in the park and the neighboring hinterlands. The hunting of other forest wildlife for food known as "bushmeat" was also frequent in this region (Hart and Hall 1996).

These censuses suggest that the population had been stable for about twenty years, judging from the total number of gorillas and the propor-

FIGURE 1. Map showing the location of Kahuzi-Biega National Park. Dotted area represents national park or forest reserve. Diagonal area represents lake.

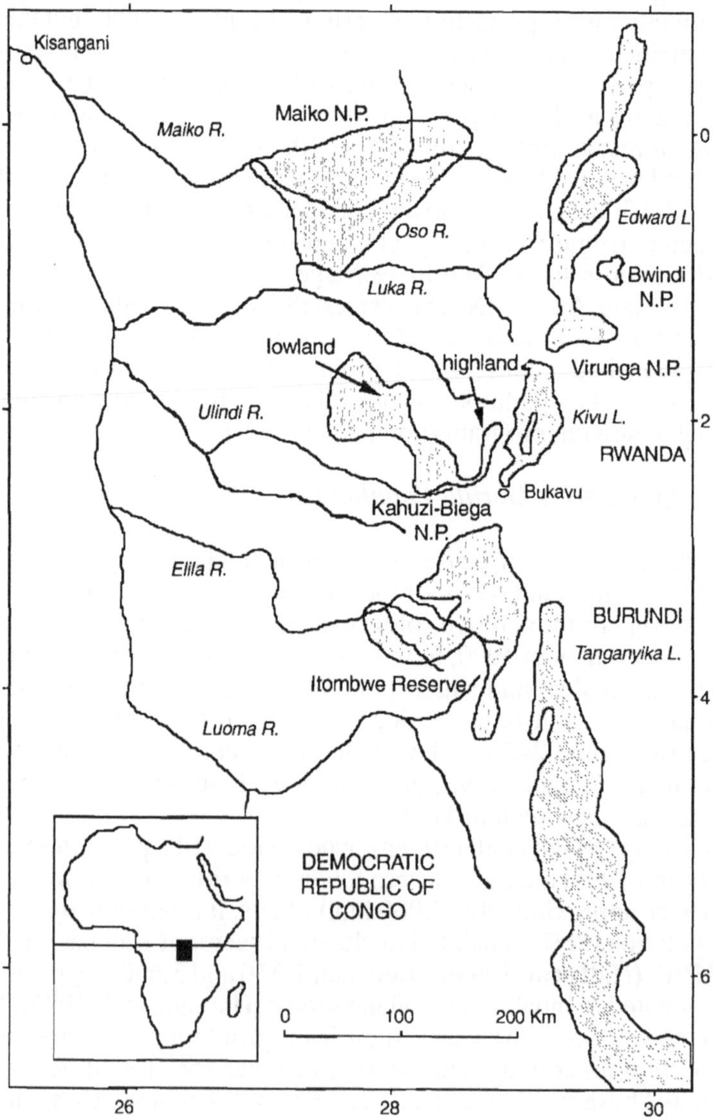

TABLE 1. History of Kahuzi-Biega National Park

1970	Park gazetted in the highland sector (600 km^2)
1972	"Gorilla Tour" initiated for public
1975	Park extended to the lowland sector (6,000 km^2)
1978	First population census of gorillas conducted (highland sector)
1980	Park inscribed in the World Heritage List
1985	Cooperative project initiated with GTZ
1990	Second population census of gorillas conducted (highland sector)
1991	Riots occurred in major cities (DRC)
1992	Establishment of a local NGO, POPOF
1994	First population census of large mammals in the lowland sector
	Influx of 450,000 Rwandan refugees around the park
1996	Third population census of gorillas (highland sector)
	First civil war begins
1997	Democratic Republic of the Congo established
	Refugees return to Rwanda
	Elephant hunting increases
1998	Second civil war begins
1999	Gorilla hunting increases
2000	Fourth population census of gorillas conducted (highland sector)

tion of infants within the population. However, the mean group size drastically decreased between the 1978 and 1990 censuses. The decrease in mean group size might have reflected a reduction in habitat quality as the population was compressed into a smaller area. In 1990 and 1996, researchers observed that the frequent poaching of other animals caused some gorillas to concentrate in the central well-protected area of the park (Yamagiwa et al. 1993; Inogwabini et al., in prep.).

The 1996 census estimated the first reliable number of elephants to be 910 individuals in the highland sector (Inogwabini et al., in prep.). However, about 94% of these elephants tended to concentrate in the well-protected central area. The densities of monkeys and antelopes were estimated to be very low, probably due to heavy poaching. More than 60% of the highland sector was found to be threatened, as determined by an analysis using GIS (Geographic Information System) (Inogwabini 1997).

However, populations of gorillas, elephant, and other large mammals in Kahuzi-Biega National Park have plummeted in the past few years. Despite major gains produced by conservation efforts during the 1980s and early 1990s, bushmeat poaching has caused a drastic decline in the population of the eastern lowland gorilla. Poaching of gorillas occurred on a massive scale in 1999 (Yamagiwa 1999). All four of the gorilla groups that had accepted tourists no longer exist, and researchers sus-

TABLE 2. Comparison of Population Sizes and Structures at the Times of Three Censuses Taken in Kahuzi-Biega National Park (Highland Sector)

	1978	1990	1996
No. of groups	14	25	25
No. of solitary males	5	9	2
Total no. of gorillas	23	258	245
Mean group size	5.6	10.8	9.7
% infant	17.0	8.4	12.7
Source:	Murnyak 1981	Yamagiwa et al. 1993	Vedder 1996 Inogwabini et al. (in prep.)

pect that about half of the highland sector's gorilla population has been killed. The current situation is one of crisis, and the species now faces extinction. Commercial and subsistence hunting has eliminated over 95% of the elephant population since the early 1990s.

The causes and motivations of the upsurge of poaching and the deterioration of the species' populations are not simple. Economic collapse, social unrest, military conflict, disruption of conservation activities, and conflicts among local people all contributed to increased poaching of elephants and gorillas. I have witnessed some of these social, political, and ecological changes while conducting research in the park (since 1978) and recently while working on a non-governmental organization (NGO) project organized by local people. The aim of this paper is to analyze the escalation of wildlife poaching in Kahuzi-Biega National Park and consider effective conservation measures to reverse current trends.

HISTORY OF CONSERVATION EFFORTS AND THE IMPACT OF THE REFUGEE CRISIS AND CIVIL WARS

Background on Conservation Efforts and the Crisis

In the 1980s, people in this region faced a severe economic crisis. The national currency was drastically devalued and the cost of living rose ten-fold, while wages had hardly doubled (Pay and Goyvaerts 2000). The payment of wages to civil workers was delayed by months, or in some cases, over a year. Despite the worsening socioeconomic situation, gorilla poaching was rare, although snares set in the park for antelope occasionally wounded gorillas. This reflected the success of cooperative efforts made by IZCN and GTZ with support from the

World Bank, the World Wide Fund for Nature (WWF), and other international NGOs. The IZCN-GTZ project helped local people by building dispensaries and schools. It also improved public health and helped lessen the workload of local women by providing water piped from natural sources (Steinhauer-Burkart et al. 1995). These successes were notable given that the region around the park has one of the highest population densities in the region (more than 300 individuals/km^2), and in recent decades the population has grown at an annual rate of 4% (Institut National de la Statistique 1984).

Gorilla tourism was well organized during this time, with visitors taken to four habituated groups. It also generated significant revenues. From 1989-93, annual revenues from gorilla viewing were about US$210,000 (Butynski and Kalina 1998). Park authorities decided to devote 40% of park revenues to park management and community development projects near the park. Consequently, the number of park guards and patrol stations were increased, and improvements were made to park facilities and equipment. The benefits derived from gorilla tourism helped boost support for gorilla conservation throughout the region.

However, following the outbreak of riots in Kinshasa, the capital of Zaire, in 1991, the number of tourists and the revenue from tourism drastically declined. After the riots, the Mobutu regime headed toward collapse. Most foreign businessmen left the country and foreign aid and cooperation were suspended. Hospitals and clinics faced severe shortages of medicines and supplies. To keep schools and colleges open, parents of students organized to pay the wages of teachers. Rather than resigning from their formal jobs, many people used their social status to supplement their income with the collection of bribes, black-marketeering, or smuggling. In the national parks, the government paid the park guards less than US$10 per month, and the payment of their salaries and those of other civil servants was usually delayed for several months. About "80% of all financial resources of the national park infrastructure were consumed in the capital city" of Kinshasa (Hart et al. 1996, 685).

The Rwandan genocide in 1994 exacerbated these problems by provoking the influx of about 450,000 refugees into the area around the park. These refugees were driven from the area in 1996 during the civil war that led to the end of the Mobutu regime in 1997. Although the new government, headed by Laurent Kabila, succeeded in controlling the rampant hyperinflation, a second civil war erupted in 1998 and the communities around the park were again shaken by the sound of gunfire.

The second civil war ushered in a period of intensive gorilla poaching during 1999 (Yamagiwa 1999). Most of the known gorilla hunting that occurred in the highland sector took place while park guards were disarmed and prohibited from patrolling the park by the rebel government of the Ressemblement Congolais pour la Démocratie (RCD). The RCD seized the Kivu region in 1998 and has controlled it since. Armed militia groups have frequently encamped in the lowland sector of the park, which has been out of the control of the RCD government, and poaching there is likely to be even more intense than in the highland sector. Theses militias included local groups known as "Mai-Mai" as well as the *interahamwe*–comprised of former Rwandan soldiers opposed to the RCD.

POPOF and Local Conservation Efforts

In 1992, as the political and economic situation in the Kivu region deteriorated, people from villages adjacent to Kahuzi-Biega National Park established the Pole Pole Foundation (POPOF). The idea for POPOF was proposed by John Kahekwa, who had experience as a guiding tourists and working with foreign film crews. POPOF's first members included other local park guides, and its initial funding came from foreign visitors. The main purpose of POPOF was to promote local conservation knowledge, improve the quality of life in the area's communities, and reduce conflicts between local people and the park. For many local residents, the park had been a longstanding source of resentment and conflict. When the national government created the park in the 1970s, it required many local villagers to abandon their lands and refrain from using the natural resources of the new reserve. They were also prohibited from shooting the elephants that frequently raided their crops. Furthermore, local villages were ordered to absorb the people who were evicted from the new reserve.

POPOF sought to promote local community development through a variety of projects that would simultaneously benefit conservation in the park and neighboring lands. These projects included the establishment of a handicraft center and a tree nursery that employed local people, including former poachers. A school for women and children was established and ecotourism was organized in and around the park. Based on these activities, POPOF solicited foreign aid for its continued operation. However, due to the increased instability caused by the civil war in neighboring Rwanda in 1993, the number of tourists drastically decreased, and consequently POPOF failed to realize its plan to become self-sustaining.

The Refugee Crisis and the Park

In 1994, some 450,000 fled the genocide in Rwanda and arrived to the area around Kahuzi-Biega National Park (Hall 1994). The sudden and massive influx of refugees disrupted local communities and severely hampered park management. The United Nation High Commission for Refugees (UNHCR) set up two sprawling refugee camps near the park that accepted 50,000 and 200,000 people, respectively (Figure 2). A large number of foreign staff from the UNHCR and international NGOs stayed near the camps or in the nearest city, Bukavu. These organizations transported large quantities of aid material into the region, revitalized the system to distribute goods, and employed many local residents.

FIGURE 2. Original sector of Kahuzi-Biega National Park. Map A shows the location of 25 groups and 9 solitary males of gorillas during the 1990 census. Map B shows the vegetation types of the park and the location of refugee camps in 1994. Dotted area: *Cyperus* swamp; Slashed area: bamboo forest; White area within the park: primary and secondary forests. Black circle: patrol station of the park.

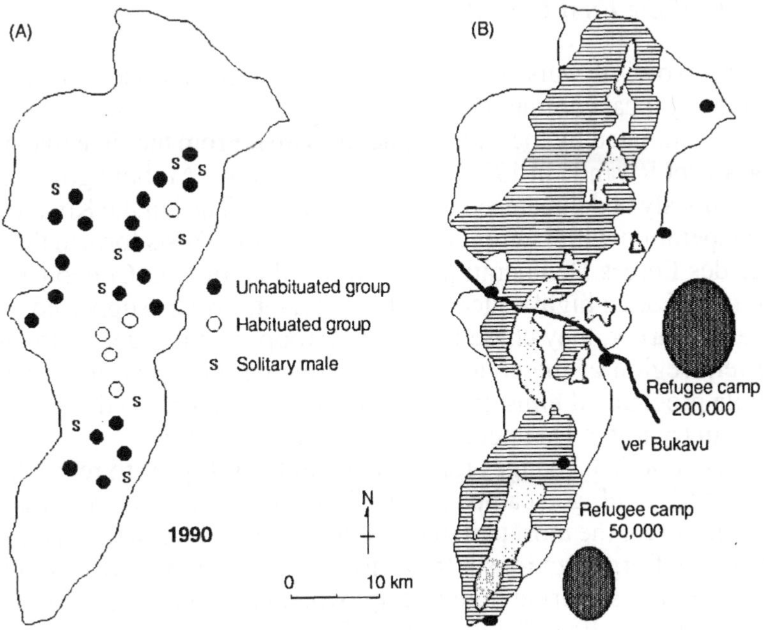

Pressure on the natural resources of Kahuzi-Biega National Park grew, and damage to the natural environment was particularly severe around the refugee camps. The UNHCR and GTZ purchased firewood from village residents to supply the camps. The enormous demand for firewood led to accelerated deforestation in and around the park. Tree felling and intensified cultivation of crops resulted in the expansion of farmland into the park. Another problem during this period was the spread of small arms. Rwandan soldiers and militias introduced many firearms into the camps and ordered some local residents to hunt elephants for the ivory trade. During this time, unemployed villagers, who were known to have hunted elephants in the park before, were put on trial for poaching. The prosecution of the villagers contributed to the gradual development of conflict between them and park employees. With support from GTZ, the park headquarters increased the number of guards and improved equipment for patrols. However, they barely succeeded in protecting the park.

EFFECTS OF CIVIL WARS ON THE PARK

Effects of the First Civil War

The two civil wars led to more intense exploitation of natural resources by local communities. The first civil war began in October 1996 and was initiated by the Banyamulenge Tutsi. From the time that genocide swept Rwanda in 1994, *interahamwe* and the Mobutu government continuously had threatened the Banyamulenge Tutsi in the Kivu region. In cooperation with Laurent Kabila's troops, the Tutsis formed the Alliance des Forces Democratiques pour la Liberation du Congo (AFDL) and organized a highly successful counter-offensive. Supported by the Rwandan army, they chased Mobutu's troops and refugees into the far western region of the country, finally seizing Kinshasa in May 1997. Kabila proclaimed himself the new president and changed the name of the country to the Democratic Republic of the Congo.

These political changes forced many of the refugees to return home. The massive refugee camps around Kahuzi-Biega National Park were closed, and for the most part, the UNHCR and the international humanitarian relief organizations left the area. Many communities around the park were left suffering high unemployment and shortages of food and other basic supplies. The departure of the humanitarian groups and the subsequent desperation of the local communities contributed to an abrupt increase in poaching within the park.

Elephants were heavily poached in the early months of the first civil war in 1996. For two months after the outbreak of the war, insecurity prohibited park guards, guides, and trackers from returning to the park headquarters and stations. The task of the park guards was made more difficult when they were disarmed by the new government's armed forces. When the guards resumed patrolling the park, they found many dead elephants. One entire group, including infants, was found together dead. Tusks were taken from each elephant, and although small pieces of meat were smoked in some places, evidence indicated that the ivory trade was the primary reason for the slaughter. Since then, elephants or their footprints have rarely been observed anywhere in the park. It is believed that nearly the entire elephant population of the park was killed.

Before the outbreak of the first civil war, gorillas were rarely hunted in the park. In 1993, unemployed villagers had killed one famous male gorilla, named Maheshe. Park guards surmised that the incident was the result of growing conflict between unemployed villagers and park employees (Park guards, pers. comm.). After the outbreak of war, two silverback (adult) males from two of the habituated groups were killed. Poachers who had disagreements with trackers employed by the park killed the first silverback in April 1997. Six months later, government soldiers on patrol in the park shot the second silverback out of fear when they encountered a habituated group. The rest of the gorilla groups remained unaffected by poachers.

The Second Civil War and Natural Resource Plunder

The Kabila government succeeded in controlling the country's hyperinflation but people in the Kivu region did not feel the benefit. The government's increasing aggression towards the Banyamulenge Tutsi provoked another civil war, which erupted in August 1998. In contrast to the previous war, this conflict was characterized by the active participation of many neighboring countries. Both the Kabila government and the rebel government in the eastern part of the country permitted foreign troops to exploit natural resources, such as diamonds, gold, cobalt, coltan, copper, and timber (Sawada 2000; Prendergast and Smock 1999). The internal conflict and the process of natural resource plunder became more complex as neighboring governments purposely drove rebel combatants from their respective civil conflicts into the DRC and continued to fight them there. This phenomenon appears to have protracted the war in the DRC.

After the outbreak of the second war, education and public works were suspended and many hospitals were closed. People suffered serious shortages of food and medicine, as armed soldiers of the rebel government restricted trade, the transport of goods, and even human movements. Many people, including thousands of refugees around Kahuzi-Biega Natural Park turned to traditional medicines, such as crude drugs made from wild plants and animals. Traditionally, the local people have depended heavily on the park's natural resources for many basic needs, including medicine (Sivha 2000). Interviews with 213 people from 25 villages show that 92.6% of 249 wild plant species they use come directly from the park. More than 70% are used for medicine.

With the renewed conflict came a new wave of intense poaching and the exploitation of mines within the park. Heavily armed *interahamwe* and Mai-Mai militias frequently entered the park and engaged in poaching. The park guards were disarmed during the first civil war, and have remained unarmed. With the renewed fighting, patrols by the park guards were suspended for seven months. They resumed at the end of March 1999. During their absence, poachers freely entered the park to shoot wild animals and set snares. The felling of trees for construction and the collection of firewood also markedly increased. At the same time, there was accelerated encroachment of slash-and burn farming along the entire border of the park. The active destruction and exploitation of the park's resources may have been partly a reflection of increased hostility felt by local people toward the park and its authorities.

GORILLA POACHING AND THE PARK'S CONSERVATION MEASURES

The State of Gorilla Groups in the Park

Until the end of July 1998, park staff had monitored five gorilla groups on a daily basis. Four groups (Mushamuka, Maheshe, Nindja, and Mubalala) were monitored for the purposes of tourism (Table 3) and another group (Ganyamulume) for the purpose of scientific research. These five groups were comprised of 94 gorillas. At the outbreak of the second civil war in August 1998, rebels made it impossible for the park staff to enter the park. The park's main entrances were closed and transport stopped between the lowland and the highland sec-

TABLE 3. Size and Composition of Habituated Gorilla Groups Before (1998) and After (1999) Large-Scale Hunting

Age/sex class Years old		Silverback Over 13	Blackback 10-12	Adult female Over 10	Subadult 7-9	Juvenile 4-6	Infant 0-3	TOTAL
Mushamuka	Before	0	1	3	1	1	2	8
Group	After	0	1	3*	1	1	0	6
Maheshe	Before	1	0	11	0	1	2	14
Group	After	0	0	0	0	0	0	0
Nindja	Before	0	0	14	2	7	2	25
Group	After	0	0	3	0	0	2	5
Mubalala	Before	1	1	13	1	2	4	22
Group	After	0	0	0	0	0	0	0
Ganyamulume	Before	1	1	12	3	4	3	24
Group	After	1*	0	12	2	5	3	23
TOTAL	Before	3	3	53	7	15	13	94
	After	1(0)	1	18(15)	3	6	5	34(30)

* Indicates the number of immigrants from unknown groups
Number in parenthesis represents original members surviving the large-scale hunting

tors. Although monitoring of the study group resumed after three months, no direct observations were made of the gorillas. Little information was available on the four habituated groups until the end of March 1999, when the park resumed monitoring of the groups and regular unarmed patrols. The following summaries describe what was learned about each gorilla group.

The Mubalala Group–The members of this group have not been seen in their former range since July 1998. In February 1999, the park staff found a large number of gorilla bones scattered throughout their known range. In some places, piles of burned bones were found. Nearby villagers also reported that dead bodies of gorillas had been transported by poachers to a village along the lowland sector of the park. It is likely that most members of the Mubalala group were killed for the bushmeat trade. In June 1999, the park staff found a fresh nest site of gorillas in the former range of the Mubalala group. They tried to contact the gorillas and confirmed that at least two females showed no fear when approached. These females may be immigrants from the habituated groups (most probably from the Mubalala group). The park named this group Mufanzala (after a tracker in the park) and started to monitor them.

The Maeshe Group–In April 1999, the park staff located and resumed monitoring the Maeshe group (23 gorillas). The group had moved to the former range of the Mubalala group, where intensive poaching was now taking place. Gunfire was frequently heard in the area. At the end of July 1999, the group disappeared and no nests were found. Within weeks, park staff found a poacher's hut and a large number of gorilla skulls. Poachers had apparently smoked gorilla meat on a fire. In September 1999, the park authorities arrested a group of poachers who possessed numerous fragments of gorilla fur, skulls, and bones. The poachers may have killed most members of the Maeshe group for bushmeat.

The Mushamuka Group–When park staff resumed monitoring and patrolling activities in April 1999, they were unable to located the Mushamuka group. It had probably disintegrated before the resumption of patrols. However, a small group consisting of a young silverback, three females, a sub-adult, and a juvenile was found in the group's former range. Subsequently, it was confirmed that the maturing silverback was Kaboko, born in the Mushamuka group in 1987. As a young gorilla he had lost his right hand to a snare. Now that he had formed his own group, he was renamed Mugaruka, after the chief of the nearby village. A sub-adult male had also lost his right hand and had his left hand disabled. The park staff continue to monitor this group.

The Nindja Group–At the beginning of April 1999, this group consisted of 19 nest-builders with five infants. Weeks later, the sound of gunfire was heard in the range of the Nindja group, and since then, the group has not been found. Some time later, a number of dead gorilla bodies were seen being carried by poachers to villages in the vicinity of the park. Poachers probably shot and killed most members of the group. In July 1999, a group of gorillas moved into the former range of the Nindja group. The park started to monitor and habituate this group. They found at least three females of the Nindja group associating with them. Apparently, other gorillas have recently joined the group. In October 1999, the total number reached 31. The group included a silverback and at least five juveniles and three infants. The silverback, Mishebere (named after a tracker in the park), had a disabled left hand (probably the result of a snare). A 3 year-old juvenile also had a disabled right hand. Mishebere and the other members of the group were habituated quickly and began to accept visitors in September 1999.

The Ganyamulume Group–Most of the gorillas in this group escaped the killing, although poachers killed a young silverback in August 1999. Thereafter, a solitary male began to associate with the group, and all of the females and immature gorillas have since remained with him as a unit. However, other neighboring groups have disappeared from this area. Four groups had ranged around the study site (about 30 km^2) until August 1998, but three were not found in October 1999. Like the habituated groups, these groups were likely hunted for bushmeat (Figure 3).

Conservation Measures Taken by the Park to Control Poaching

In September 1999, the park authorities summoned sixty-seven suspected poachers and asked them about their activities. They claimed they had hunted due to hunger. They were subsequently promised a pardon for any poaching that had occurred during the war. Most of them had recently hunted both elephants and gorillas, some very close to the park headquarters. The park authorities decided to employ forty of them to assist with patrols and the tracking of gorillas. Their activities included dismantling snares and poachers' huts. Guides and trackers paid daily visits to the Mugaruka, Mishebere, and Mufanzala groups, and a

FIGURE 3. Gorilla groups confirmed by us before and after the period of intensive poaching. Map A shows the location of 25 groups and 9 solitary males of gorillas during the 1990 census. Map B shows the location of gorilla groups confirmed in October 1999.

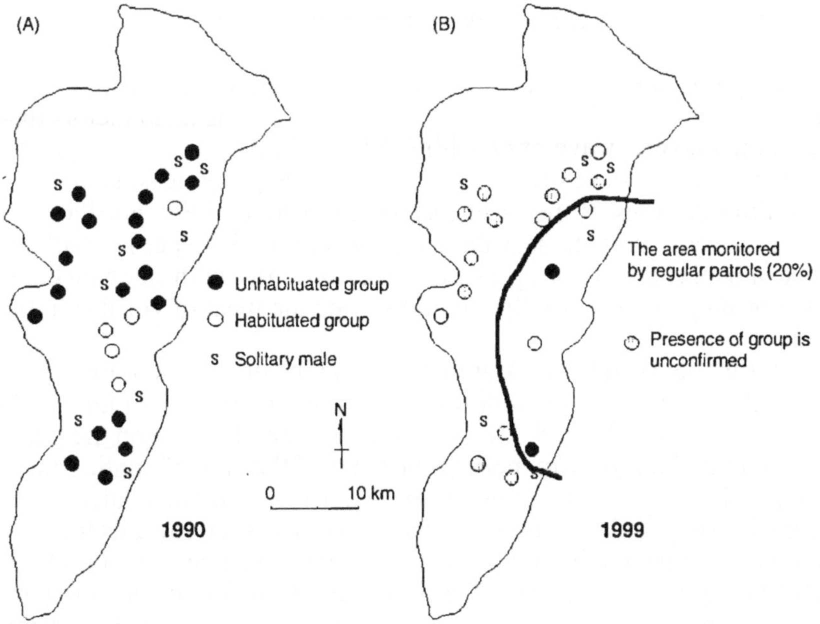

few guides remained to observe each group from 6:00am to 6:00pm. Since the park contracted the poachers, poaching activity has significantly decreased and few gorillas have been lost in the habituated groups.

The fourth and most recent census of the gorillas was carried out between June and August, 2000 by the Congolese researchers supported by Wildlife Conservation Society (WCS), The Dian Fossey Gorilla Foundation (DFGF), and GTZ. They found 126 gorillas in thirteen groups and four solitary males (Kanyunyi Basabose, pers. comm.). These results show that half of the gorillas have been lost between 1996 and 2000, although the mean group size (9.6) and the proportion of infants (9.3%) have not drastically changed. Elephants were not found, but a few fresh trails were seen, indicating that probably only a few elephants survived. By 2000, the park had lost more than 95% of the elephant population estimated in the 1996 census.

At the end of 2000, the park staff's frequent patrols were keeping 20% of the highland sector safe for gorillas and elephants. The rest of the park is not monitored by guards and seems to be frequented by poachers. The lowland sector is beyond the park authorities' control, and the situation there may be worse.

Factors Influencing the Poaching Crisis in the Park

Starvation and the spread of small arms among the local people, combined with the breakdown of park protection were the main factors that led to the massive wave of poaching. The main purposes of the gorilla poachers were to stave off their own hunger and generate some income by selling portions of the meat locally. In October 1999, gorilla meat sold in several markets near the park for only US$.25 per kg–half the price of beef (Park guards, pers. comm.). In contrast with elephant hunting, gorilla poachers usually cut and smoke the meat of gorillas in the park.

Conflict between the park authorities and the unemployed local people could be an important additional factor that predisposes individuals to hunt gorillas in the park. The poaching initially focused on the gorillas habituated for tourism, perhaps because of the ease of hunting these groups. However, the targeting of these gorillas could be related to the strong hostility and resentment that some villagers felt towards the park and the local people it employed. Interviews conducted by POPOF of local villagers and poachers confirmed that hostility toward the park was growing among unemployed people in the area. The sudden termination of hunting after the employment of the former poachers in September 1999 may support this hypothesis.

CONCLUSIONS AND RECOMMENDATIONS

Large-scale hunting of gorillas in Kahuzi-Biega National Park was one of the tragic consequences of the social, political, and economic instability that developed in the region with the collapse of the Mobutu regime, the refugee crisis, and the subsequent civil wars. The most effective solution to the current crisis is to establish a lasting peace through constructive negotiations among all stakeholders. However, an end to the war in the DRC would not mean an end to the conservation problems in Kahuzi-Biega National Park. Communities around the park still lack sufficient supplies of food and other basic materials, and insecurity remains a severe problem.

The conflicts between the park authorities and local people may continue to be a factor that motivates desperate people to poach gorillas. Certainly more information could be gathered to understand and thus more effectively address the underlining factors that drive people to engage in poaching. Nonetheless, it is important to undertake efforts that engender a broader acceptance of the park among local people. Steps must be taken to create mutually beneficial relationships between the park and local residents if local attitudes towards the park and its gorillas are to improve. The employment of former poachers by the park is one example of the kind of efforts that can be made to improve local conservation knowledge and build trust between the park authorities and local people.

Even during the present crisis, there are some conservation activities that can be undertaken and planning for future conservation efforts should continue. The International Gorilla Conservation Program (IGCP) along with GTZ, WCS, WWF, DFGF, IGCP, and other agencies and conservation organizations are now working to identify and implement appropriate measures to maximize the impact of conservation during and after periods of armed conflict. Extensive research has been conducted on the bushmeat trade, and an international campaign to stop the trade has been organized (Bowen-Jones 1998). A major conference on apes held in May 2000 in Chicago focused the attention of researchers on the plight of Africa's gorillas and the urgent need for greater conservation efforts on their behalf. Since then, many researchers have called on the governments of countries with great ape populations to stop the bushmeat slaughter of apes and the destruction of their habitat. The International Primatological Society decided to promote a declaration of World Heritage status for the great apes at the society's 18th Congress in Adelaide, Australia, in January 2001. These actions aim to improve the protection of the great apes by lobbying for better conservation legislation and enforcement in countries with ape populations.

However, a bottom-up approach to conservation is also very important during times of armed conflict (Hart and Hall 1996; Mubalama 1999). With weak enforcement of rules and regulations and national institutions in disarray, local people in the DRC have increasingly made decisions according to short-term personal or local interests. Park authorities are powerless to stop local people from using protected areas for agricultural production and the provision of other necessities. Under such desperate conditions, NGOs such as POPOF play an important role by reinforcing conservation knowledge among local people and providing alternatives to destructive activities. Foreign countries and international NGOs should support their efforts to protect communities and endangered wildlife from the ravages of war. The following strategies are recommended:

1. Facilitate peaceful negotiations among all political forces
2. Support improvements to the park's facilities, the infrastructure around the park, and the equipment used to patrol the forest
3. Establish a conservation and education center near the park for local people and tourists
4. Reconsider compensation for the people who were evicted when Kahuzi-Biega National Park was established, and consider compensating the villages that bore the burden of receiving them
5. Establish regulations and enforcement capacity to prevent the bushmeat trade and dismantle the poacher-trader network
6. Provide conservation education and training for soldiers, including instruction on how to behave in encounters with wildlife
7. Increase the participation of local people in the park management and tourism
8. Develop and implement a plan for the optimal distribution of profits from tourism among local communities
9. Support efforts to organize tourism and promote its acceptance in the area around the park
10. Establish effective methods for monitoring wildlife in the park

Congolese researchers working in cooperation with various international NGOs helped initiate some of these conservation activities by conducting the 2000 gorilla census in the highland sector of the park. However, the census results indicate that park management strategies, including the promotion of gorilla tourism, should be reevaluated. Habituation of gorillas and promotion of tourism may not be the best solution for conservation of gorillas and for development of local communities. It may lead to increased rates of infection and disease among gorillas and may ultimately weaken the viability of the gorilla population (Butynski and Kalina 1998). The influx of refugees and the prolonged human settlement inside Kahuzi-Biega National Park may have already increased the risk of disease transmission from humans to wildlife (Mubalama 1999). As mentioned earlier in this report, it is plausible that the habituation of gorillas facilitated the rapid poaching of certain gorillas. Care should also be taken to make sure that the benefits from tourism are distributed equitably among the local people who bear the greatest burdens of the park. Otherwise, conflict among local people may increase and lead to further exploitation and abuse of resources in the park. The above strategies should be conducted in the near future to help improve local knowledge and understanding of the eastern lowland gorillas as an important part of local, national, and world heritage.

REFERENCES

Bowen-Jones, E., 1998. A Review of the Commercial Bushmeat Trade with Emphasis on Central-West Africa and the Great Apes. A Report of Ape Alliance, "The African Bushmeat Trade–A Recipe for Extinction." pp. 8-47.

Butynski, T.M. and J. Kalina, 1998. Gorilla tourism: a critical look. pp. 294-313 in E.J. Milner-Gulland and R. Mace (eds.), Conservation of Biological Resources, Blackwell Science Ltd., Oxford.

Corbet, G.B., 1967. The nomenclature of the eastern lowland gorilla. Nature, 215: 1171-2.

Goodall, A.G. and C.P. Groves, 1977. The conservation of the eastern gorillas. pp. 599-637 in H.S.H. Prince Rainier III and G.H. Bourne (eds.), Primate Conservation, Academic Press, New York.

Groves, C.P., 1967. Ecology and taxonomy of the gorilla. Nature, 213: 890-3.

Hall, J.S., 1994. Counting gorillas in Zaire's Kahuzi-Biega NP. African Wildlife Update, November-December: 6.

Hall, J.S., Saltonstall, K., Inogwabini, B-I. and I. Omari, 1998a. Distribution, abundance, and conservation status of Grauer's gorilla (*Gorilla gorilla graueri*). Oryx, 32: 122-30.

Hall, J.S., White, L.T., Inogwabini, B-I., Omari, I., Morland, H.M., Williamson, E.A., Saltonstall, K., Walsh, P., Sikubwabo, C., Bonny, N., Kaleme, P.K., Vedder, A. and K. Freeman, 1998b. Survey of Grauer's gorillas (*Gorilla gorilla graueri*) and eastern chimpanzees (*Pan troglodytes schweinfurthii*) in the Kahuzi-Biega National Park lowland sector and adjacent forest in eastern Democratic Republic of Congo. International Journal of Primatology, 19: 207-35.

Hilton-Taylor, C. (compiler), 2000. 2000 IUCN Red List of Threatened Species. IUCN, Gland, Switzerland and Cambridge, UK.

Harcourt, A.H. and D. Fossey, 1981. The Virunga gorillas: decline of an 'island' population. African Journal of Ecology, 19: 83-97.

Hart, T.B., Hart, J.A. and J.S. Hall, 1996. Conservation in the declining nation state: a view from eastern Zaire. Conservation Biology, 10: 685-6.

Inogwabini, B-I., 1997. Using GIS to determine habitat use by large mammals and to define sensitive areas of Kahuzi-Biega National Park, Eastern Congo. Msc. Thesis, University of Kent, England.

Inogwabini, B-I., Hall, J.S., Vedder, A., Curran, B., Yamagiwa, J. and K. Basabose, (in preparation). Pre-war conservation status of large mammals in the mountain sector of Kahuzi-Biega National Park, Democratic Republic of Congo. African Journal of Ecology (submitting).

Mankoto, M.O., 1988. La gestion du Parc National du Kahuzi-Biega (Zaire). Cahiers d'Ethologie Appliquee, 8: 447-50.

Mubalama, L., 1999. A view from eastern Democratic Republic of Congo. pp. 5-10 in L. Naughton-Treves (eds.), Fighting in the Forest: Biodiversity Conservation Amidst Violent Conflict, Conservation and Development Forum, Gainesville.

Murnyak, D.F., 1981. Censusing the gorillas in Kahuzi-Biega National Park, Biological Conservation, 21: 163-76.

Pay, E. and D. Goyvaerts, 2000. Belgium, the Congo, Zaire and Congo: a short history of a very shaky relationship. pp. 1-36 in D. Goyvaerts (ed.), Conflict and Ethnicity in Central Africa, Institute for the Study of Languages and Cultures of Asia and Africa, Tokyo University of Foreign Studies.

Prendergast, J. and D. Smock, 1999. Putting Humpty Dumpty Together: Reconstructing Peace in Congo. http://www.marecine.com/NCN.hml.

Ruvolo, M., Pan, D., Zehr, S., Goldberg, T., Disotell, T.R. and M. von Dornum, 1994. Gene trees and hominoid phylogeny. Proceedings of the National Academy of Sciences, USA, 91: 8,900-8,904.

Sawada, M., 2000. Remarks on the war in the Democratic Republic of Congo. pp. 37-58 in D. Goyvaerts (ed.), Conflict and Ethnicity in Central Africa, Institute for the Study of Language and Cultures of Asia and Africa, Tokyo University of Foreign Studies.

Sivha, M., 2000. Conservation of resources in Kahuzi-Biega. Gorilla Journal, 19: 6-7.

Steinhauer-Burkart, B., Muhlenberg, M. and J. Slowik, 1995. Kahuzi-Biega National Park. The IZCN/GTZ-Project Integrated Nature Conservation in East-Zaire.

Vedder, A., 1996. Rapport préliminaire du projet de gorilles: secteur originale du Parc National de Kahuzi-Biega. Typescript, Wildlife Conservation Society, New York.

Von Richter, W., 1991. Problems and limitations of nature conservation in developing countries: a case study in Zaire. pp. 185-94 in W. Erdelen, N. Ishwaran and P. Muller (eds.), Tropical Ecosystems, Margraf Scientific Books, Weikersheim.

Yamagiwa, J., 1999. Slaughter of gorillas in the Kahuzi-Biega Park. Gorilla Journal, 19: 4-6.

Yamagiwa, J., Mwanza, N., Yumoto, Y. and T. Maruhashi, 1994. Seasonal change in the composition of the diet of eastern lowland gorillas. Primates, 35: 1-14.

Yamagiwa, J., Mwanza, N., Spangenberg, A., Maruhashi, T., Yumoto, T., Fischer, A. and B. Steinhauer-Burkart, 1993. A census of the eastern lowland gorillas Gorilla gorilla graueri in Kahuzi-Biega National Park with reference to mountain gorillas *G. g. beringei* in the Virunga region, Zaire. Biological Conservation, 64: 83-9.

The Chainsaw and the Gun: The Role of the Military in Deforesting Indonesia

Charles Victor Barber
Kirk Talbott

SUMMARY. Since the mid-1960s, the Indonesian military has played an active role in the exploitation of the nation's forest resources. This role is best understood within the historical context of the military's pervasive and institutionalized involvement in the social, political, and economic affairs of the nation. The military's abuse of power and its direct involvement in the exploitation of the nation's forests were partly the result of the doctrine of *dwi-fungsi* (dual function) implemented under the regime of President Suharto (1966-98). Forest resources played an important role in the Suharto regime's plans for economic development and the extension of its political control across the nation. The military's involvement included: (1) direct investment in forest industries; (2) facilitation of contracts and concessions through influence peddling and the forceful repression of communities and other interests opposed to logging; and (3) participation in many aspects of the illegal timber economy. The post-Suharto era has been one marked by political upheaval, economic crisis, and efforts to decentralize political power. Since Suharto's

---

Charles Victor Barber is Senior Associate, World Resources Institute, 10 G Street N.E. Suite 800, Washington, DC 20002.

Kirk Talbott is Vice President for Asia/Pacific, Conservation International, 1919 M Street, NW, Suite 600, Washington, DC 20036.

The authors would like to thank J. Brunner, O. Lynch, F. Hobart, S. Price, and the anonymous reviewers for their comments on early drafts of this paper.

[Haworth co-indexing entry note]: "The Chainsaw and the Gun: The Role of the Military in Deforesting Indonesia." Barber, Charles Victor, and Kirk Talbott. Co-published simultaneously in *Journal of Sustainable Forestry* (Food Product Press, an imprint of The Haworth Press, Inc.) Vol. 16, No. 3/4, 2003, pp. 137-166; and: *War and Tropical Forests: Conservation in Areas of Armed Conflict* (ed: Steven V. Price) Food Products Press, an imprint of The Haworth Press, Inc., 2003, pp. 137-166. Single or multiple copies of this article are available for a fee from The Haworth Document Delivery Service [1-800-HAWORTH, 9:00 a.m. - 5:00 p.m. (EST). E-mail address: getinfo@haworthpressinc.com].

fall and during the subsequent period of *reformasi*, greater attention has been paid to forest issues and the military's role in logging. However, the military's longstanding involvement in illegal logging and unsustainable forest practices will likely continue if the broader movements to reform civil society are stymied. Conservationists should link their efforts with a range of development and advocacy organizations to support the broader reform movement. If the country's unique and globally significant forest patrimony is to be conserved, the military's substantial role in deforestation must be acknowledged and promptly addressed.

KEYWORDS. Forest policy, forest conservation, Indonesia, biodiversity, military, deforestation, timber, civil society, corruption, TNI

INTRODUCTION

Over the past three decades, the Indonesian Military (Tentara Nasional Indonesia, TNI) has played a pervasive and multifaceted role in the deforestation in Indonesia. Under the "New Order" regime of President Suharto (1966-98), TNI operated at the very center of power, doing Suharto's political and economic bidding and profiting handsomely in return. Suharto's regime enshrined a "dual function" (*dwi fungsi*) for the military, under which TNI had a prominent role within the nation's sociopolitical affairs as well as a traditional role as a national security force. Freed from most restraints on its power, TNI charted a course marked by systematic human rights abuses, pervasive meddling in civilian political matters from Jakarta to the village level, and extensive involvement in numerous sectors of the economy.

The exploitation of Indonesia's vast forest resources was an important part of the New Order's economic strategy, and TNI was involved in at least three ways. First, TNI has been a direct investor in logging operations, most notably in the 1970s. Second, local TNI units have frequently provided security for forest-related business enterprises that have come into conflict with local communities over forest resources. Third, military officials, particularly at the provincial and local levels, have actively facilitated (through extortion and bribery) and participated in illegal logging and other illegal uses of natural resources (such as poaching).

This paper reviews the deterioration of Indonesia's forests over the past decade and analyzes the role of TNI in that process. It goes on to review the major political and economic changes since the overthrow of Suharto in May 1998. His fall from power and the subsequent growth of political and economic reform movements (*reformasi*) initiated a tumultuous series of events that are reshaping the role of TNI within Indonesia's rapidly evolving political and economic system. These epochal changes and the dramatic *reformasi* period have given strength to those calling for reform of Indonesia's forest policies and practices. However, forest policy reform will be difficult unless fundamental changes are made in the longstanding role of the military vis-à-vis the civilian government, civil society, and the economic system. The military's substantial role in deforestation is a crucial linkage in the larger mosaic of crisis and opportunity facing Indonesia today and must be promptly addressed if the country's unique and globally important forests are to be conserved.

INDONESIA'S FORESTS UNDER THE SUHARTO REGIME– A LEGACY OF DESTRUCTION AND CONFLICT

Forest Resources of Indonesia

Indonesia possesses the greatest remaining expanse of tropical rainforest in Southeast Asia, and is one of the planet's most important repositories of forest biodiversity. These forests are an important source of livelihood for millions of forest-dependent people, including many indigenous forest-dwelling peoples with long-standing customary (*adat*) traditions of forest resource management. The forests also play an important role in the national economy.

Approximately three-fourths of Indonesia's land area (see Figure 1), more than 140 million ha, is legally classified as "forest land." However, much of this area is already deforested or in varying states of degradation. A 1999 estimate, based on 1997 satellite imagery, concluded that about 95.8 million ha still had some form of forest cover, but did not shed light on the condition of those forest areas (The World Bank 2000b). Indeed, a 2000 analysis carried out by Indonesia's Ministry of Forestry and Estate Crops concluded that only about 36 million ha of primary forest remained, while an additional 14 million ha consisted of logged-over forest still in medium or good condition.

FIGURE 1. Map of Indonesia.

Biologically, these remaining forests are extremely diverse. While Indonesia occupies only 1.3% of the world's land area, it possesses about 10% of the world's flowering plant species, 12% of all mammals, 17% of all reptile and amphibian species, and 17% of all birds (National Development of Planning Agency [BAPPENAS] 1993, 1-2). Forest habitats include evergreen lowland dipterocarp forests in Sumatra and Kalimantan, seasonal monsoon forests in the eastern islands of Nusa Tenggara, non-dipterocarp lowland forests in Papua (formerly Irian Jaya), and the world's largest areas of mangrove forest. Many islands have been isolated for millennia, and levels of endemism are high. Indonesia's forests, particularly in the "Sundaland Hotspot" (which includes the islands of Sumatra and Borneo), are indisputably among the Earth's most biologically diverse and endangered terrestrial eco-regions (Mittermeier et al. 1999; Jepson et al. 2001).

Products from Indonesia's forests constitute a significant part of the national economy. During the 1990s, forest products, on average, contributed about 6-7% of Gross Domestic Product (GDP) and 20% of foreign exchange earnings, with forest product revenues in 1998 totaling US$8.5 billion, ranking second only to oil (The World Bank 2000a). Indonesia's forests also yield many non-timber forest products (NTFPs). With an export value of US$360 million in 1994, rattan canes are the most valuable NTFP (De Beer and McDermott 1996, 74). Exports of wildlife and plants for the 1999-2000 fiscal year was more than US$1.5 billion according to the Ministry of Forestry and Plantations (2000). These statistics do not take into account the immeasurable value of the food, clothing, shelter, and medicine that indigenous populations derive from the forests or the critical environmental services (such as carbon storage and watershed protection) provided by forests.

A large but undetermined number of forest-dwelling or forest-dependent communities live in or adjacent to Indonesia's forests. Estimates made over the past several decades have varied wildly from 1.5 to 65 million people, depending on which definitions were used and which policy agenda was at stake (Zerner 1992). In 1997, the logging and forest products sectors directly employed over 183,000 people, but as many as 30 million people depend directly on the forestry sector for their livelihoods (Ministry of Forestry and Plantations 2000). Many of these forest-dwellers live by long-sustainable "portfolio" economic strategies which include combinations of activities such as shifting cultivation, fishing, hunting, the gathering of forests products (such as rattan, honey, and various resins), and the cultivation of tree crops such as rubber (Capistrano and Marten 1986).

The value of these forest products is poorly appreciated, as it is not reflected in formal market transactions. It is partly obscured by the political and economic marginalization of rural and forest dependent communities. As in the case of many developing regions, the marginalization of forest communities in Indonesia is exacerbated by forest laws and policies largely predicated on the state's usurpation of traditional, community-based tenurial rights (Lynch and Talbott 1995).

Deforestation

Until mid-1999, Indonesia's annual deforestation rate was variously estimated to be between 0.6 and 1.2 million ha (Sunderlin and Resosudarmo 1996). A forest mapping project carried out in 1999 by the government (with support from the World Bank) determined that the average annual deforestation rate for 1986-97 was around 1.7 million ha (see Table 1). Based on 1997 satellite imagery, the study revealed that Sumatra was the most heavily affected. Some 30% of the huge island's forest cover had vanished during those 11 years (The World Bank 2000b). Over the past 15 years, Indonesia appears to have lost 24 million ha of forest, an area roughly the size of Laos or the United Kingdom. If trends continue, all non-swampy forestry will be lost in Sumatra by 2005 and in Kalimantan by 2010 (The World Bank 2000b). Deforestation has also brought an increase in poaching, and some of Indonesia's best-known large mammal species, such as the Sumatran tiger (*Panthera tigris* ssp.

TABLE 1. Forest Cover and Deforestation in Indonesia, 1985-97

	Forest Cover 1985		**Forest Cover 1997**		**Deforestation**		
	Total Cover 1985 (ha)	% Total Land Area	Total Cover 1997 (ha)	% Total Land Area	Decrease 1985-97 (ha)	% Loss	Loss ha/year
Sumatra	23,324,000	49	16,632,000	35	6,691,000	29	558,000
Kalimantan	39,986,000	75	31,512,000	60	8,474,000	21	706,000
Sulawesi	11,269,000	61	9,000,000	49	2,269,000	20	189,000
Maluku *	6,348,000	81	[> 5,544,000]	?	> 800,000	13	67,000
Irian Jaya	34,958,000	84	33,160,000	81	1,798,000	5	150,000
TOTAL	**115,885,000**	**68.5**	**95,848,000**	**57**	**20,505,000**	**17**	**1,709,000**

* Data for Maluku are preliminary

Source: World Bank. 2000. Deforestation in Indonesia: A Review of the Situation in 1999 (Draft Report). Jakarta.

sumatrae) and the Sumatran orangutan (*Pongo abelii*), are facing an extremely high risk of extinction (Hilton-Taylor 2000).

The Dynamics of Deforestation in the Suharto Era

These alarming rates of deforestation have their roots in the forestry policies and practices of the Suharto era. Outside Java, there was little commercial exploitation of forest resources until the late 1960s, and forests were for the most part governed by local systems of *adat* law and resource management. However, under the Suharto regime this situation rapidly changed, as a constitutional mandate of state "control" over forests was interpreted to mean that the state "owned" the nation's forestlands (Peluso 1993).

Under the Basic Forestry Law passed in 1967, the Ministry of Forestry (at that time actually a department within the Ministry of Agriculture) was empowered to "determine and regulate legal relations between individuals or corporate bodies and forests, and deal with legal activities related to forests." In accordance with the new law and its numerous implementing regulations, certain lands were classified as "Forest Area" by ministerial decree. At one point, this designation was applied to 144.5 million ha, fully 76% of the country's land area, and included many non-forested areas. These lands were further sub-classified for production, conversion to non-forest land uses, or some type of protection. Under the province-by-province process of Consensus Forest Land Use Planning (TGHK) undertaken in the 1980s, the government-controlled Forest Area in each province was divided into these various categories and recorded on official maps. The TGHK forest land-use designations have been revised numerous times since the mid-1980s, but still represent the basic legal framework for forestland use (Table 2).

The most important consequence of this system was the rapid parceling out of more than 500 renewable 20-year logging concessions to private sector firms and state-owned enterprises during the 1970s and 1980s. Official figures state that these concessions generated 612 million m^3 of logs between 1970 and 1999, but some industry analysts have argued that actual removals were approximately twice this volume (Barr, in press).

In the early 1980s, Indonesia banned the export of raw logs, catalyzing the rapid growth of the plywood processing industry as well as rapid concentration of ownership in the forestry sector. By 1990, Indonesia had become one of the world's major plywood producers. The fifteen largest plywood groups owned 65 of the nation's 132 plywood firms, accounting for 54% of the industry's overall production capacity. To supply their operations, they also controlled 151 HPH (Hak Pengusahaan Hutan) timber concessions covering some 18 million ha, nearly one-third of the 55 million ha of concessions allocated at the time (Barr 1998). By contrast, only 17.3 million ha of the country's forests were included in conservation areas, with an additional 30.1 million ha designated as "protection forest" (generally water catchments and steeply sloping areas).

During the 1990s, additional areas were allocated for industrial timber plantations (4.7 million ha) and large-scale plantation development, primarily for oil palm (3 million ha). Some of the plantations were former logging concessions that had been degraded by poor logging practices (see, Casson 2000; Potter and Lee 1998). Nearly 2 million ha were also allocated for clear-cutting and use by the government's "transmigration" program. The transmigration program, which officially ended in 2000, relocated rural people from the over-crowded volcanic islands of Java and Bali to the forested "outer islands" primarily Sumatra, Kalimantan, Sulawesi, and Irian Jaya (western New Guinea, renamed Papua in 1999).

Logging concessions, transmigration, and plantation development are among the officially sanctioned activities that have propelled deforestation in Indonesia, but illegal logging is also widespread and systematic in many areas. In fact, Indonesia's timber economy can now be said to be largely an illegal, underground economy. The *Jakarta Post*, 3 July 1996, reported that illegal timber brokers flourish throughout the country and provide logs to processors who cannot obtain adequate supplies

TABLE 2. Permanent Forest Land Categories and Proportion Holding Forest Cover

Functional Category	Designated Forest Land Area, 1999 (ha)	Actual Extent of Forest Cover (ha)	% of Designated Area with Actual Forest Cover
Protection Forest	30,100,000	24,100,000	80
Park and Reserve Forest	17,300,000	14,900,000	86
Production Forest	30,600,000	24,700,000	81
Limited Production Forest	31,000,000	25,300,000	83
TOTAL	109,000,000	89,000,000	82

Source: J. Fox, M. Casson and G. Applegate, Forest Use Policies and Strategies in Indonesia: A Need for Change. Jakarta. Paper prepared for the World Bank. May, 2000.

legally. Concession roads often provide access to illegal loggers who are encouraged by the lack of meaningful access controls by either the logging firms or local forestry officials. According to the Ministry of Forestry and Estate Crops, illegal logging has become so well-organized, backed, and extensively networked that "it is bold enough to resist, threaten, and in fact physically tyrannize forestry law enforcement authorities." Consequently, illegal logging occurs "in concessions, areas of forest slated for conversion, and in conservation and protection forests" (Ministry of Forestry and Estate Crops 2000, 13).

According to a study by the Indonesia-United Kingdom Tropical Forest Management Programme (1999), annual illegal removals are thought to be in the range of 30 million m^3, exceeding legal cutting and thus supplying the majority of the country's timber (see Table 3). A senior official of the Ministry of Forestry and Estate Crops presented an even grimmer view, revealing that the ministry's 1998 data showed that legal log production in that year was only about 21 million m^3 (down from 30 million m^3 in 1997). Meanwhile, illegal logging increased to 57 million m^3 and accounted for 70% of all wood production for the year. In an article published in the *International Herald Tribune*, 1 February 2000, the same senior official asserted that, "the wood-processing industry has been allowed to expand without reference to the available supply of timber, resulting in vast over-capacity. The shortfall in the official timber supply is being met largely by illegal logging, which has reached epidemic proportions."

TABLE 3. Timber Supply in Indonesia, 1998: Estimated Contributions of Legal and Illegal Logging

Number of timber concessions	464 units
Area of timber concessions	51,251,052 ha
1. Estimated legal production from concessions	15,769,385 m^3
2. Estimated legal production from forest land conversion	10,162,080 m^3
3. TOTAL LEGAL PRODUCTION OF TIMBER (1+2)	25,931,465 m^3
4. Total amount of timber used by licensed mills	46,587,681 m^3
5. Use of illegally-logged timber by licensed mills (4-3)	20,656,216 m^3
6. Estimated illegal logging for pulp and paper industry and domestic use	11,943,784 m^3
TOTAL ILLEGAL TIMBER PRODUCTION	**32,600,000 m^3**

Source: D.W. Brown, 1999. Addicted to Rent: Corporate and Spatial Distribution of Forest Resources in Indonesia. Jakarta: DFID/ITFMP.

The *Jakarta Post*, 15 May 2000, reported that corruption among civilian and military officials, many of whom are closely involved in illegal cutting and marketing, is pervasive. A June 2000 government analysis officially recognized the main actors in illegal logging as "(a) laborers from communities in the forest areas and also many who are brought there from other areas; (b) investors, including traders, concession holders, or holders of legal timber cutting permits (IPK), and buyers of illegal timber from processing industries; and (c) government officials (both civilian and military), law enforcement personnel, and certain legislators" (Ministry of Forestry and Estate Corps 2000, 13-14).

According to the *Indonesian Observer*, 20 June 2000, official involvement in illegal logging has become so blatant and widespread that provincial legislators in Sumatra's Jambi province felt obliged to make a public appeal to military, police, and justice officials to stop supporting illegal logging operations. *Asia Pulse/Antara*, 23 June 2000, reported that even the Indonesian Wood Panel Association (Apkindo) complained that illegal sources from Sumatra and Kalimantan were supplying at least 1 million m^3 of the 7 million m^3 sold to the Chinese market. In response to the problem, the international aid agencies and lending institutions grouped in the Consultative Group on Indonesia (CGI) have issued a number of warnings that continued aid to the forestry sector is contingent on more effective action to eradicate illegal logging.

The Suharto regime's forest policy benefited a relatively small circle of well-connected officials and their business cronies, but it was animated by broader political motives as well (Barber, Johnson, and Hafild 1994). With some three-fourths of the nation defined as state-controlled "forest areas," the application of the New Order forest policies to the hinterlands was central to the regime's efforts to extend its political authority over the remoter parts of the sprawling archipelago. Political and economic power in Indonesia is overwhelmingly concentrated on the island of Java, which makes up only 7% of the country's land area, but is home to some two-thirds of the population and the majority of economic activity. Thus, in addition to the forests' value as a source of state revenue and patronage, "forest policy" became an important arena for the New Order's program of economic development, political control, and ideological transformation (Barber 1997; Guinness 1994; Dove 1988). In this way, forests served a "dual function" for the New Order regime.

THE MILITARY, POLITICS, AND BUSINESS IN THE SUHARTO ERA

The Military and Politics

The extensive involvement of TNI in politics during the Suharto era provided the framework for its role in the exploitation of the nation's forests. The roots of TNI's dominant political role lie in the birth of independent Indonesia in the late 1940s and 1950s. TNI coalesced out of the armed guerilla groups that fought the Dutch colonial government from the end of the Second World War until 1949. With independence won, TNI argued that its role in winning independence had been larger than that of the civilian diplomats. Furthermore, it argued that it was a popular institution that had arisen "from the people" (International Institute for Democracy and Electoral Assistance [International IDEA] 2000). In doing so, the TNI began to establish a vision of itself as a legitimate holder of political as well as military power.

As early as 1947, the military had begun to set up a parallel system of government in which each civilian official, from provincial governors down to district supervisors, would be paired with a corresponding military officer (Schwartz 1994). TNI's political powers were reduced during a brief and chaotic period of parliamentary democracy during the 1950s. However, they were revived with the advent of founding President Sukarno's authoritarian "Guided Democracy" system (1957-65) and the restoration of the 1945 Constitution (which provides the legal basis for military participation in politics). By the 1950's, the TNI had become "the main power broker in Indonesia" and its ascent culminated in 1996 with General Suharto's controversial takeover of executive authority (International IDEA 2000, 87).

TNI's political role is enshrined in the doctrine of *dwi fungsi* (dual function) under which it has "a defense and security function, as an instrument of the state, and a political function, as a guardian of the state and participant in the political process" (Lowry 1996, xxi). This doctrine was codified by Indonesia's highest legislative body, the People's Consultative Assembly (MPR), in the Defense Act of 1982. It states that TNI is "a social force" that will "guarantee the success of the national struggle in development and raising the people's standard of living." The law goes on to say that TNI will promote development by "participating in decision-making in state business and improving *Pancasila* [for explanation, see, Lowry 1996, 178] democracy and constitutionality according to the 1945 Constitution in all aspects of development" (Law No. 20/1982, Article 28).

The pervasive role of the military in the nation's political life is manifested in TNI's "territorial command" structure and its "sociopolitical" institutions and functions. TNI is divided along conventional lines into an army, navy, and air force. The army consists of approximately 240,000 personnel and is organized into two major groups: central and territorial forces. About 85% of the army's forces are distributed between eleven "territorial commands" across the nation (Lowry 1999). These territorial forces are comprised of regular servicemen and women rather than reserves or a militia. Lowry explains that this kind of deployment is due to "the unfortunate fact is that regular troops are required to meet the constant demand for internal security operations, the imperatives of regime maintenance, and to prevent the chances of popular military training being misused" (1996, 93-94).

For most of the Suharto era, the military was utilized to enforce internal obedience to the regime's ideology and policies. This was accomplished through a variety of means. First, sociopolitical (*Sospol*) staff sections were incorporated into the TNI structure from top to bottom. Their role was to ensure that policies formulated by parliamentary representatives or bureaucrats were consistent with regime objectives and policies, and to monitor all political developments in society. Together with the TNI intelligence agency and the sociopolitical directorate in the Ministry of Home Affairs, the *Sospol* staff monitored and regulated all political and social organizations, and ensured the election of "appropriate" candidates to national and local offices. Second, TNI service personnel were appointed directly to 20% of the seats of the legislature at national, provincial, and district levels. Finally, seconded TNI personnel were regularly posted to jobs ranging from cabinet minister to village heads, and to positions in state-owned economic enterprises. TNI did not fill all such positions, but was well represented in them, especially in those agencies that it deemed politically or economically strategic (Lowry 1996). All of these political tools were utilized by the military during the Suharto era to impose a broad definition of "security" that covered "all aspects of politics, ideology, economics, society and culture, as well as military matters" (International IDEA 2000, 86).

The Military's Role in Business

Like its political role, TNI's involvement in businesses reaches back to the revolution, when many army and militia units had to find their own sources of income. Most of the enterprises were disbanded in the immediate post-revolutionary period but were slowly revived on a small scale after 1952 in response to drastic budget cuts and rising insurgencies. As Dutch holdings were nationalized and handed over to the military in late 1957, the army began to run large-scale businesses, often in collaboration with Chinese businessmen known as *cukong*. Many of these nationalized businesses, such as plantations, sugar mills, and hotels, declined rapidly with the rest of the national economy in the early part of the 1960s.

The first decade of the New Order regime revitalized TNI's business prospects. The army was politically dominant, oil and timber booms were underway, and the economy was reviving. Suharto, who himself had a long history of business enterprises while an active TNI officer, used the awarding of contracts to cement the loyalties of various TNI factions. TNI businesses became involved in a variety of projects ranging from construction of buildings, roads, and ports to distribution of agricultural inputs, logging (discussed below), and numerous aspects of the oil industry. However, by the late 1970s, Suharto had firmly established his power and distanced himself from the military businesses. These operations generally went into decline, but TNI's links with smaller Chinese and indigenous businesses in the provinces continue to be important (Lowry 1996).

The military's business activities today can be grouped into four categories: (1) cooperatives operated to improve the soldiers' welfare; (2) business ventures by military units, such as the use of TNI ships and vehicles to transport material and people for a fee; (3) the investments and business activities of military-run foundations (*yayasan*); and (4) the informal "facilitation" of private business ventures in exchange for fees (essentially extortion or bribe-taking). The total value of TNI's businesses is not known with any accuracy, but the *Inter Press Service*, 20 November 1998, reported that a 1998 study by the Indonesian Institute of Science (LIPI) estimated total TNI business assets to be at least US$8 billion.

These business activities have become an institutionalized part of the TNI budget, since Indonesian defense spending is far higher than that declared in the official government budget. Indeed, there is currently no other way to fund the basic needs of service personnel, since regular salaries do not come close to meeting these needs. One Indonesian economist has suggested that the declared defense budget probably accounts for only 25% of actual defense spending, with the rest coming from the TNI's various economic activities. Some of these funds are siphoned off by particularly well placed individuals, others are reinvested into companies, and some become extra-budgetary income for the military (McCullouch 2000).

THE MILITARY AND THE FORESTS IN THE SUHARTO ERA

The military's influence on Indonesia's forests during the Suharto era can be grouped into three types of activities. During the first decade of the New Order regime, TNI became directly involved in the ownership of millions of hectares of logging concessions. Secondly, local TNI units and personnel have often acted as enforcers for logging and plantation firms that have run into resistance from local communities. Third, TNI personnel have been deeply involved in the illegal timber economy.

The Military as Timber Tycoons

As noted above, large-scale logging of the commercially-valuable dipterocarp forests on Indonesia's outer islands, especially Kalimantan and Sumatra, commenced soon after the transfer of power from Sukarno to Suharto in 1966. Logging provided solutions to two of the most pressing problems of the new regime. First, the economic chaos of the mid-1960s had reduced government revenues to a trickle, making it difficult to fund government operations and salaries, and this had led to widespread corruption and absenteeism in the armed forces and the bureaucracy. A quick infusion of cash into the state apparatus was badly needed. Second, Suharto's hold on power was quite tenuous at the outset, and he saw the granting of privately-operated logging concessions as one way to cultivate and consolidate loyalties among both the military and the civil bureaucracy. The concession system allowed Suharto to "transfer substantial economic rents–that is, profits above and beyond normal rates of return–to favored clients at all levels of the state apparatus" (Barr 1998, 4).

Between 1967 and 1970, Soedjarwo, Suharto's hand-picked director general of forestry, handed out 519 logging concessions covering over 53 million ha to private investors, and an additional 4 million ha were distributed between three state-owned forestry enterprises. During this period, ventures involving foreign companies committed US$424 million, more than 80% of investment in the sector. The military played a direct role in this timber boom, forging partnerships with some of the largest investors (Barr 1998). It was observed that, "military-owned holding companies, cooperative enterprises, foundations, and pension funds, representing the particular interests of both individual officers and whole commands, frequently acted as 'silent partners' for foreign logging companies" (Barr 1998, 6).

Due to the lack of transparency in the military-run businesses and foundations, the full extent of TNI's direct participation in the timber boom of the 1970s is not known, but it was certainly extensive. The activities of the military-owned PT Tri Usaha Bakti (Truba) provide one documented example of this phase of TNI involvement in logging. Truba was formed in 1968 through a merger of 40 businesses established by army officers in the mid-1960s. All of its shareholders were senior officers in the Ministry of Defense. By the late 1970s, Truba reportedly held interests in 14 logging ventures, many of which were set up by military officers who had served in areas with prime commercial timber stocks.

One prominent venture was Truba's partnership with the US-based timber giant Weyerhaeuser in the International Timber Corporation of Indonesia (ITCI), which obtained a 601,000 ha concession in East Kalimantan in 1971. While the initial capital investment in this venture was US$32 million, Truba contributed only US$150,000 (0.5%) but nonetheless held a 35% interest. Truba's involvement in ITCI exemplified military enterprises entering such partnerships, in that its "contribution to the venture was essentially the concession itself, which Weyerhaeuser would not have been able to obtain on its own" (Barr 1998, 6).

A ban on log exports, phased in during the early 1980s, caused many TNI businesses to either sell out their concessions or affiliate with the larger conglomerates who progressively came to dominate the timber/plywood sector by the end of the that decade, as noted above. TNI is still thought to have extensive interests in (legal) logging, but it does not play the dominant, direct role in the industry that it did in the 1970s. However, military personnel continued to play an important role as facilitators of business deals in both the logging and plantation sectors, selling their influence to help investors obtain access to favored forest areas and ensure local government cooperation.

The Military as Paid Enforcers

Local army and police units also appear to have "moonlighted" as paid enforcers during the Suharto era. They used TNI's monopoly on state-sanctioned violence to suppress local resistance to takeovers of community lands by commercial ventures. New Order forest policy essentially entailed the takeover of millions of hectares of forestlands that were formerly under the customary (*adat*) control and management of millions of people living in traditional forest-dependent communities. Inevitably, this wholesale usurpation led to numerous conflicts with local communities. Where such conflicts got out of hand, logging and plantation companies would pay local police and military personnel to intimidate local resisters into acquiescence. The techniques of intimidation ranged from verbal intimidation to intensive interrogation, torture, and murder. Such practices appear to have been particularly widespread in resource-rich Aceh province, where TNI has been fighting a separatist movement for decades. Although poorly documented, numerous anecdotal accounts by victims and witnesses indicate that such practices were widespread in the Suharto era (for details, see, Barber 1997, Barber forthcoming).

The Military Role in Illegal Logging

As noted previously, illegal logging appears to have increased in the 1990s. The military has played a significant role in this illegal trade, ranging from acting as hired protection for illegal logging operations to transporting illegally cut logs on military vehicles. Examples are numerous, although most documented examples date from the post-Suharto period, when a newly freed press began to report on what had been common knowledge for years.

The *Jakarta Post*, 7 November 2000, reported that one forestry official in Lampung province (Sumatra) admitted that illegal logging was common in the area and often involved both military and civilian officials. In Central Kalimantan's Tanjung Puting National Park, an important reserve for the endangered orangutan, blatant and systematic illegal logging is carried out by local timber barons with the apparent blessing of local police and military. When a barge carrying 1,500 illegal logs cut from the park was seized by the authorities in early June 1999, it was auctioned off two weeks later to the very timber baron responsible for logging the park (Environmental Investigation Agency and Telapak Indonesia 1999). According to the *Jakarta Post*, 14 July 2000, widespread illegal logging on the Indonesian-Malaysian border in Kalimantan was a "joint venture" involving protection by military personnel from both countries. In Riau province (Sumatra), illegal logging is pervasive in protected forests, and according to one report in *Detikworld*, 3 September 2000, military and civilian officials "are involved and back up the theft." In Aceh province (Sumatra), the military's intimate involvement in illegal logging includes the regular use of military vehicles to transport illegal timber to market (Barber and Nababan, unpublished report). So extensive is the role of military (and civilian) authorities that a 2000 Ministry of Forestry and Estate Crops report noted that, "government officials (both civilian and military), law enforcement personnel, and certain legislators" are key actors (Ministry of Forestry and Estate Crops 2000, 13-14).

Detikworld, 3 April 2001, reported that Suripto, the former secretary general of the Ministry of Forestry and Estate Crops, acknowledged that, "the military HQ is in the pockets of the wood smuggling Mafia. The Mafia's action is a long chain . . . [and] there is backing the police, military members, customs officers and forestry people and also members of the legislature." Indicating that evidence gathered during his tenure pointed to similar problems in Sumatra and Papua as well, he went on to say that, "What is going by land is clearly with the backing of policy and ground force military commanders. What is going by sea is backed by the Navy; only the Air Force is not mixed up in this wood smuggling." In March 2001, President Wahid fired Suripto, who was widely respected as one of the few forestry officials seriously pursuing corruption and malpractice in the sector. Wahid had accused Suripto of various acts of political disloyalty and even "treason," and according to the *Indonesian Observer*, 28 March 2001, Suripto is currently suing the president for slander.

THE MILITARY, SOCIETY, AND FORESTS IN THE POST-SUHARTO ERA

The complex web of economic and political interests described above did not vanish when Suharto was driven from office in May 1998. Indonesia's political system has since been in a period of chaotic and violent transition. The role of TNI is in flux as well. Political upheaval and economic crisis have considerably diminished the military's power. Continuing revelations about its abuses during and after the Suharto era have left it disgraced it in the eyes of many. Nonetheless, it remains the most powerful single political actor in the country. As Indonesia gropes towards a functional and democratic system of governance, transforming the role of the military remains one of the greatest challenges. Similarly, any transition to sustainable and equitable management of the nation's forests–or at least an end to the spiral of deforestation–also requires a fundamental change in the relationships among civilian government institutions and the military. To understand the importance of this daunting challenge, it is necessary to review the economic and political events that have shaken Indonesia since 1997.

The Economic Crisis, the Fall of Suharto, and the Rise of Reformasi

The East Asian economic crisis that began with the devaluation of the Thai baht in July 1997 affected Indonesia more severely than any other country in the region. The national economy remains moribund and is essentially staying afloat with its oil revenues and an International Monetary Fund (IMF)-led bailout program. Most of the country's banks, and many other important economic players, are bankrupt, the value of the Indonesian rupiah remains low, and unemployment and inflation are high. While political unrest had been growing throughout 1996 and 1997, the economic crisis was a major catalyst for the crescendo of opposition and violence that drove President Suharto from office in May 1998.

The legitimacy of Suharto's 32-year rule was largely dependent on the delivery of continued economic growth. In exchange for this prosperity, many elements of society were willing to tolerate rampant corruption, regular abuses of human rights, and the absence of democratic political processes (for details on Suharto era corruption, see "Audit Implicates" 2000; "Audit Shock" 2000). With the economy spiraling into depression, support for the aging president–even among many who had long served in his government and military–evaporated in an explosion of student-led protests and riots (for detailed accounts, see Forrester and May 1999; "Indonesia After" 1998; "Indonesia May" 1998; Scott 1998).

The demise of the New Order regime left Indonesia in a state of political limbo under the transitional government of Suharto protégé President B. J. Habibie. Parliamentary elections, the first relatively free and fair elections in 44 years, were held in June 1999. Four months later, the MPR elected Abdurrahman Wahid president and Megawati Sukarnoputri vice-president.

The centrifugal tendencies inherent in such a large, multi-ethnic archipelagic nation have been unleashed since the end of the New Order. East Timor–invaded and occupied by Indonesia in 1975–was finally given its independence after a violently-contested referendum process supervised by the United Nations in mid-1999 (United Nations Office of the High Commissioner for Human Rights [UNHCR] 2000). Long-simmering separatist movements in the resource-rich provinces of Aceh and Papua (formerly called Irian Jaya) have been reinvigorated, and other provinces (such as oil-rich Riau, in Sumatra) have begun to talk about independence as well.

Ethnic and religious killing, looting, ordinary violent street crime, and brutal vigilantism have exploded all over Indonesia since 1998. According to the *Far Eastern Economic Review*, 7 July 2000, Moslem-Christian violence in the eastern province of Maluku has taken thousands of lives. The *Jakarta Post*, 7 July 2000 and 15 February 2000, reported similar savagery in parts of Kalimantan and Sulawesi. According to *Agence France-Presse*, 20 June 2000, the government estimated that there were more than 765,000 "internal refugees" fleeing these various conflicts. Urban crimes rates have soared, and the *Far Eastern Economic Review*, 13 July 2000, reported that hundreds of suspected street criminals have been beaten and burned to death in public places under a savage form of "street justice."

The government is moving rapidly towards a new system of "regional autonomy." This is being done partly in response to the separatist movements, but also in reaction to more widespread resentment of the centralized governance of the Suharto era. However, the provincial and district governments that will benefit from this sweeping decentralization are, for the most part, completely lacking in the capacities needed to govern effectively. Many are still run by entrenched and corrupt holdovers from the Suharto era (see, "Workshop Questions" 2000; "Logical Flaws," 2000).

Furthermore, the nation's legal system is widely considered corrupt and inefficient. In fact, the *Indonesian Observer*, 5 April 2000, reported that senior officials even debated "importing" Dutch judges on the assumption that no honest ones could be found in Indonesia (for analysis of corruption in the Indonesia legal system, see, Lindsey 2000; "Matter of Law," 2000; "Reform of the Legal," 2000). The disgrace into which the legal system has fallen has damaged the credibility of the numerous on-going investigations into corruption and human rights abuses of the Suharto era.

Virtually all elements of the political spectrum have adopted the rhetoric of *reformasi*: democratization of politics, respect for human rights, and the elimination of "corruption, collusion, and nepotism" (generally termed KKN, from *korupsi, kolusi, dan nepotisme*). *Reformasi*, however, means very different things to different people. For the many holdovers from the old regime who are still in power or are biding their time, it means removing the rough edges and the most blatant corruption from the current system but generally continuing business as usual. For students and other more radical reformers, it means nothing less than the complete burial of the New Order regime and the creation of a democratic political system. In Aceh and Papua, provinces with long-standing separatist movements, many view *reformasi* as an opportunity to gain at least a greater measure of autonomy, if not independence, from the central government. In almost every province, *reformasi* is equated with greater decentralization of political power and increased local access to the profits of natural resource exploitation.

The Changing Fortunes and Role of the Military Since 1998

The events of 1997-99 fundamentally weakened TNI's role and reputation, and the economic crisis crippled its capacities. In the aftermath of Suharto's fall, calls came for the reform of TNI. Proponents of reform pressed for the TNI to abandon the *dwi fungsi* doctrine, completely withdraw from politics, and abolish the system of territorial commands (Lowry 1999). The clamor for reform were fueled by a growing stream of revelations about past and ongoing human rights abuses in Aceh, Papua, and East Timor.

It became increasingly evident that TNI was unable to contain the growing communal violence that had wracked the country since 1998. *Agence France-Presse*, 12 July 2000, reported that the minister of defense admitted that the police and military were unable to maintain order in the country. In some cases, TNI factions were accused of inciting or directly contributing to such clashes, especially in Maluku (Saunders 2000). TNI's reputation declined further when it supported the activities of murderous vigilante gangs before East Timor's independence referendum in September 1999 (UNHCR 2000). According to an article in the *Australian*, 9 October 2000, military scandals and controversies contributed to the widespread public perception that the chain of command had broken down and that many elements in the army were provoking civil violence to undermine democracy and justify the need for military security.

Continual revelations of corruption by military commanders and institutions associated with the military have steadily eroded TNI's reputation (for details, see "Kostrad's 'Money Dredger' " 2000; "Indonesia Army" 2000; Borchier 1999). Indonesian public esteem for TNI has been at an all-time low. An August 1998 poll in major cities, conducted by a respected research institute, found that 46.5% of respondents felt that TNI did not work in the interests of society and 74.3% believed that the military should withdraw from politics. The survey also revealed a widespread perception that TNI routinely sided with business interests in industrial and land disputes. A Jakarta survey the following month by the *Kompas* newspaper revealed and 81% of the 1,500 middle-class respondents were against active armed forces officers serving in civilian posts, even at the lowest levels of the bureaucracy (Borchier 1999).

During his first months in office in late 1999, President Wahid quickly removed a number of Suharto loyalists and other hard-liners from top military posts. The most notable action was the firing of the TNI commander, General Wiranto, after he was found complicit in the subversion and violence that accompanied East Timor's independence referendum in September 1999. Responding to its changed political position, TNI enacted some apparent reforms in 1999 and early 2000. The *Jakarta Post*, 9 October 2000, reported that the TNI announced a revised doctrine in April 2000. According to the new doctrine, TNI affirms that it obeys and supports the concept of civilian supremacy over the military. That same year, the police force was separated from the military command, and TNI's sociopolitical role was adjusted so that all armed forces personnel seconded to non-military positions would be retired from duty (Lowry 1999). Wahid also made progress by appointing the first civilian defense minister in 40 years, but five active military men and one retired officer filled other cabinet positions.

Superficially at least, there seemed to be hope that TNI would play a constructive role in Indonesia's transition to democracy, or at least not actively sabotage the transition. These hopes were largely dashed in the second half of 2000, when conservative elements in the military essentially halted the process of reform and engaged President Wahid in a contest of wills. In August, TNI convinced legislators in a special session of the MPR to allow the military to retain a block of seats until 2009, five years longer than Wahid and many others had wanted. TNI also resisted calls for more transparent disclosure of the military's business ventures.

In the words of a senior western diplomat–cited in the *Washington Post*, 5 November 2000–"one area where Gus Dur [President Wahid's nickname] made significant inroads with reform was with the military, and now that is being completely reversed." The *International Herald Tribune*, 10 October 2000, cited another western official who said that, "the army is the rotten part of the Indonesian military. The army has retrenched to protect its money, power and position. That's what it's really all about." A review of the situation published by the *Far Eastern Economic Review*, 28 December 2000, concluded that President Wahid "is being forced to cede decision-making authority over Aceh and Irian Jay . . . to an increasingly assertive army . . . Meanwhile, military reform is at a standstill." One western diplomat cited in the report observed that, "the authority of the central government is so weakened that we're going to see a lot of local politics and provincial power plays. The military has positioned itself to have little warlords . . . Wahid is pushing buttons and nobody is responding. People are doing whatever they want to do."

Forest Policy Reform and the Military

The hardening of TNI against reform portends little action on its part to slow Indonesia's galloping rate of deforestation in the near term. What extent will TNI slow or sabotage ongoing efforts to reform forest policy and forestry practice? The answer will depend upon the extent to which counter-balancing institutions of government, civil society, and the private sector can strengthen their hand to implement forest policy reform, and more generally, develop the institutions of a more democratic and civil polity.

Forest policy reform has been hotly debated since Suharto's resignation in mid-1998. Demand for reform stems from five principal sources. First, forest policy reform is viewed as a fundamental part of the larger movement to eradicate the "KKN" that characterized the Suharto regime. Second, the IMF-led economic assistance program for Indonesia–that was initiated in late 1997 in response to the East Asian economic crisis–contains a number of forest policy reform conditionalities. Third, the growing movement for greater regional autonomy in Indonesia's provinces implies a significant shift in the balance of power between the central government and provincial and local government units with respect to natural resources, including forests. This is particularly true in those provinces that are rich in natural resources. Fourth, the advent of a more democratic political environment has given non-governmental organizations (NGOs) and the press greater political space to call for forest policy reforms. NGOs and others now regularly make public statements about forest policy that would have resulted in a prison sentence or worse during the Suharto years.

Finally, many local and indigenous communities that hold decades-old grievances against the central government and private forestry companies are not waiting for the policymakers' reforms. They are taking matters in their own hands. For example, in an article published by the *Jakarta Post*, 18 March 2000, the Association of Indonesian Forest Concessionaires (APHI) claimed that 50 timber companies–which control about 10 million ha of logging concessions in Papua (Irian Jaya), Kalimantan, and Sulawesi–had stopped their logging activities due to growing conflicts with local residents. The APHI explained that these local people not only claimed ownership of the firms' concessions but they had also frequently threatened the companies' workers.

The reform agenda is long and complex, and a thorough analysis is beyond the scope of this paper (for details, see Barber and Schweithelm 2000). There is no complete agreement on the agenda, but most reformists agree on these priorities:

- Reform the policies and practices of concession logging; cancel concessions in violation of the law or acquired through corrupt means; and broaden forest utilization concepts and policies to include multiple uses and a variety of users–including a strong emphasis on community-based forest management
- Carry out accurate inventories of remaining forest resources; grant clear legal protection as permanent forest estate to all remaining forested areas; and strengthen protection of forest areas containing the highest-priority biodiversity
- Recognize and legally protect forest ownership and use by indigenous and forest-dependent communities and assist them in the sustainable forest management
- Combat illegal logging through strengthened government and citizen monitoring and effective law enforcement

With significant forces resisting such changes, identifying these priority reforms has proven easier than implementing them. How is the military likely to react to forest reform efforts? Where military units or officials are involved in illegal logging, they will surely oppose any attempts to reduce this lucrative source of income. Military businesses and foundations will also resist any efforts to end their control of logging concessions. It is possible that they could make at least cosmetic reforms with respect to harvesting practices. A similar dynamic is likely to prevail where military interests and officers are involved in the plantation sector. It is also plausible that the military would resist the legal recognition of customary land and resource rights in areas where such recognition would diminish its access to valuable forest resources.

While many vested logging interests would oppose reform, the military is unique in Indonesia given that it possesses a legal monopoly on the use of force. A corrupt local civilian official or timber baron who resist forestry reforms will likely be much easier to displace than entrenched military interests. With budgets declining and conflicts between logging and plantation operators and communities on the rise, local military units are likely to increase their longstanding practice of hiring out personnel as enforcers to intimidate protesters.

Regional Autonomy and Forest Reform–A Downward Spiral?

Indonesia's rapid movement towards greater "regional autonomy" is an unpredictable factor in the movement for forest policy reform, and further complicates predictions about what role the military will play. The far reaching Law on Regional Government (Law No. 22 of 1999) replaces a 1974 act on regional government and a 1979 act on local government. It shifts the balance of governmental power from Jakarta to local government units (provinces and districts) and restructures government institutions all the way down to the village level. With respect to forests, the most important provision is probably Article 10(1). It states that "local government has the authority to manage natural resources occurring in its jurisdiction and shall be responsible to secure environmental sustainability in accordance with laws and regulations." It also grants power to the heads of local government units to pass regulations (on approval of the local House of Representatives) and decrees, as long as they do not conflict with higher laws and regulations or compromise the public interest.

What this means for the future of forest policy and management is likely to vary from region to region. In the *Jakarta Post*, 23 December 2000, the previous minister of forestry and plantations–sacked by President Wahid in March 2001–stated that, "the implementation of the regional autonomy law in the forestry sector has not reflected the commitment to the principle of sustainable forestry management." He went on to predict that conditions would worsen if "regional administrations treat the forests as a source of revenue." The former minister's position must be viewed rather skeptically, however, since his ministry has always treated the forests simply as a source of revenue.

A recent study of natural resource management on Sumatra's rainforest frontier describes a breakdown in the rule of law following decentralization (McCarthy 2000). It exposes the complex web of collusion and patronage networks that has been evolving for years but now appears increasingly intractable. In describing the vicious circle (*lingkaran setan*–literally a "devil's circle") of localized corruption, one businessman said, "Everyone at the local level obtains payments . . . The police or army set up posts and every truck must pay, or people are held until payments are made. Payments are then made up and down the chain of command" (McCarthy 2000, 5-6).

After the fall of Suharto there had been a resurgence of localized protest against outside timber interests with several examples of district-level civil disobedience against illegal logging and the army and police activity associated with it. The balance of power has, however, since moved back toward the logging networks. The only major factor that now appears to slow the "wild logging" in the region has been the decreasing viability of profitable local timber extraction in the wake of widespread deforestation across the island of Sumatra and other regions of Indonesia.

Robert Lowry, a long-time specialist on the Indonesian military, argues that regional autonomy will likely result in high levels of localized military-bureaucratic corruption. In the *Jakarta Post*, 21 December 2000, Lowry asserted that, "regional autonomy will present immense difficulties for the military and police mainly because they have not been reformed, and their failings will probably be more readily exposed by local politicians. There will also be tensions resulting from conflicting legislation and regulations. The embryonic nature of institutional development will probably see local politicians, bureaucrats, military, and police reach mutually beneficial arrangements in the interregnum before reform bites."

If both the former minister and Lowry are correct, regional autonomy will produce a decentralization of forestry sector corruption in which localized military powers will be important players. This would substantially frustrate forest policy reforms articulated at the center and lead to continuing loss of forests. As violent conflicts over forest resources continue, it is possible that the military will fall back on what it knows how to do best. The *Jakarta Post*, 20 December 2000, cited one analyst who, referring to Aceh and Papua provinces, observed that, "economic exploitation aside, violence is seen as the identity of the state that [these] regions are supposed to belong to."

CONCLUSION

It seems clear that the Indonesian military will not soon accept civilian supremacy, dismantle or relinquish control of its corrupt business enterprises, or renounce its practices of "internal security" terror exercised through its "territorial command" structure. Therefore, TNI's longstanding negative impacts on Indonesia's forests are likely to continue, albeit in a "decentralized" manner. What could change this depressing scenario?

The most promising route to military reform lies in strengthening the institutions of civilian government and civil society. TNI has justified its dominant role in politics and its heavy-handed territorial security apparatus as necessary in such a vast, multi-ethnic, and volatile nation. However, as Lowry notes, much of this volatility "has arisen from the failure to develop legitimate political structure and norms and address the political grievances and aspirations of the people. Moreover, no amount of military rule, no matter how well intentioned, is a substitute for responsive local and regional government within an appropriate national structure" (Lowry 1999, 12).

Thus, for the tide of forest destruction to be reversed, it is essential to build a local as well as national political environment that is conducive to establishing the rule of law and a system of governance that promotes long-term management of natural resources. Most Indonesian advocates of forest conservation recognized these issues long ago and have worked in dedicated opposition to the Suharto regime and the inefficient institutions that it left behind. In several instances, conservation and environmental organizations have been at the forefront of the movement to rebuild civil society in Indonesia and have a long track record of working closely and effectively with human rights, health, and other organizations. Conservation organizations must continue to link their efforts with the larger community–including reform-minded leaders in the military–working to rebuild Indonesia's civil society.

What are the implications for those whose primary concern is conserving Indonesia's forests and ensuring dignity and equity for the millions of indigenous and local people who depend on them? In his case study of the "wild logging" in northern Sumatra, McCarthy emphasizes the need to offer local communities and politicians viable economic alternatives to logging. He also underscores the importance of mobilizing public opinion and political consensus around long-term forest management and conservation objectives. According to him, the future of forests hinges on the active support of senior district officials who are perhaps the only ones who can reign in local office holders and the army and police units involved in illegal logging (McCarthy 2000).

Other countries in the region such as the Philippines and Thailand have seen significant reform in their militaries and their distancing from engagement in logging and other inappropriate activities. Even in Cambodia, where the military has been thoroughly involved in illegal logging for the last decade, progress has been achieved (Global Witness 2001; Talbott 1998). Strong and consistent pressure from multinational donors, the introduction of an independent observer–the British NGO Global Witness, and the use of a forest crime case-tracking system have made it increasingly difficult for the Cambodian government to tolerate the military's illegal operations without jeopardizing vital macroeconomic support. Progress has been possible because conservation groups, donor agencies, and committed government officials have directly addressed the military threat. Whether or not this approach would be effective in Indonesia remains to be seen. Nevertheless, Cambodia's recent history suggests that the international community retains significant leverage even in apparently anarchic conditions, and that the military is vulnerable to a concerted campaign of bad press, donor involvement, and public pressure.

Likewise, TNI is neither monolithic nor immune to change. The public's resentment after decades of meddling by the military in the nation's politics and business affairs has caused TNI's command to rethink its role and position in society. In fact, according to the *Far East Economic Review*, 31 May 2001, the military has demonstrated a careful, non-assertive role in Indonesia's increasing tumultuous policies in 2000-01. A new generation of officers and the recent flurry of political change–including the dramatic events in East Timor–have opened the door to fundamental reform and a reorientation of the armed forces. If a wide range of international and national organizations work together with Indonesia's government and civil society leaders to encourage these reforms, the future for Indonesia's dwindling forest patrimony may improve.

This paper, it is hoped, has convinced the reader that issues such as civilian control of the military are not as par removed from forest conservation as some might initially think. The imperative of political change sweeping across Indonesia coupled with the accelerating pace of deforestation demands effective reform of the military's role in the forest sector. These challenges require a convergence of efforts by conservation, human rights, and development organizations, as well as public international institutions, to strengthen political reform and expose and root out corruption. If international conservationists are to be credible and effective in protecting Indonesia's forests, they must forge new partnerships and work cooperatively to address the crucial nexus between forest conservation, democracy, and reform of the military in Indonesia.

REFERENCES

Audit Implicates Top Brass of Suharto Regime. 2000, July 5. Business Times (Singapore).

Audit Shock: Jakarta's Missing Billions. 2000, July 18. Straits Times (Singapore).

Barber, C.V. 1997. Environmental Scarcities, State Capacity, Civil Violence: The Case of Indonesia. American Academy of Arts and Sciences and University College, Toronto, Cambridge, MA.

Barber, C.V. and A. Nababan. unpublished report. Eye of the Tiger: Conservation Policy and Politics on Sumatra's Rainforest Frontier. World Resources Institute, Washington, DC.

Barber, C.V., N. Johnson, and E. Hafild. 1994. Breaking the Logjam: Obstacles to Forest Policy Reform in Indonesia and the United States. World Resources Institute, Washington, DC.

Barber, C.V. and J. Schweithelm. 2000. Trial By Fire: Forest Fires and Forestry Policy in Indonesia's Era of Crisis and Reform. World Resources Institute, Washington, DC.

Barr, C.M. in press. Will HPH Reform Lead to Sustainable Forest Management?: Questioning the Assumptions of the "Sustainable Logging" Paradigm in Indonesia. In C.J.P. Colfer and I.A.P. Resosudarmo (eds.), Which Way Forward? Forests, Policy and People in Indonesia. Resources for the Future, Washington, DC.

Barr, C.M. 1998. Bob Hassan, The Rise of Apkindo, and the Shifting Dynamics of Control in Indonesia's Timber Sector. Indonesia, 65 (April). Cornell University Southeast Asia Program, Ithaca.

Borchier, D. 1999. Skeletons, Vigilantes and the Armed Forces' Fall From Grace. Pp. 149-172 in N.A. Budiman, B. Hatley and D. Kingsbury (eds.). Reformasi: Crisis and Change in Indonesia. Monash Asia Institute, Clayton, Australia.

Brunner, J., C. Elkin, and K. Talbott. 1998. Logging Burma's Frontier Forests: Resources and the Regime. World Resources Institute, Washington, DC.

Capistano, A.D. and G.G. Marten. 1986. Agriculture in Southeast Asia. pp. 6-19 in G.G. Marten (eds.) Traditional Agriculture in Southeast Asia. A Human Ecology Perspective. Westview Press, Boulder.

Casson, A. 2000. The Hesitant Boom: Indonesia's Oil Palm Sub-Sector in an Era of Economic Crisis and Political Change. Center for International Forestry Research (CIFOR), Bogor, Indonesia.

De Beer, J.H. and M.J. McDermott. 1996. The Economic Value of Non-Timber Forest Products in Southeast Asia. 2d ed. Netherlands Committee for IUCN, Amsterdam.

Dove, M.R. 1988. Introduction: Traditional Culture and Development in Contemporary Indonesia. pp. 1-40 in M.R. Dove (ed.) The Real and Imagined Role of Culture in Development. Case Studies From Indonesia. University of Hawaii Press, Honolulu.

Environmental Investigation Agency and Telapak Indonesia. 1999. The Final Cut: Illegal Logging in Indonesia's Orangutan Parks. Environmental Investigation Agency, London.

Forrester G. and R.J. May. (eds.). 1999. The Fall of Suharto. Select Books. (Published in association with Research School of Pacific and Asian Studies, Australian National University), Singapore.

Government of Indonesia [GOI]. 1991. Indonesia Forestry Action Programme (Vol II). Jakarta.
Guinness, P. 1994. Local society and culture. pp. 267-304 in H. Hill (ed.) Indonesia's New Order. The Dynamics of Socio-Economic Transformation. Allen & Unwin, St. Leonards, Australia.
Hilton-Taylor, C. (compiler) 2000. 2000 IUCN Red List of Threatened Species. IUCN, Gland, Switzerland.
Indonesia After Suharto. 1998. Far Eastern Economic Review. 4 June.
Indonesia Army Printed Cash to Fund Subversion. 2000. The Age. 10 October.
Indonesia's May Revolution. 1998. Far Eastern Economic Review. 28 May.
Indonesia-UK Tropical Forest Management Programme, 1999. A Draft Position Paper on Threats to Sustainable Forest Management in Indonesia: Roundwood Supply and Demand and Illegal Logging. Report No. PFM/EC/99/01. Jakarta.
IDEA (International Institute for Democracy and Electoral Assistance). 2000. Democratization in Indonesia. An Assessment. International Institute for Democracy and Electoral Assistance (IDEA), Stockholm.
Jepson, P., J.K. Jarvie, K. Mackinnon, and K.A. Monk. 2001. The End for Indonesia's Lowland Forests. Science 292 (4):859.
Kostrad's 'Money Dredger' Springs a Leak. 2000. Tempo. 7-13 August.
Lindsey, T. 2000. Black Letter, Black Market and Bad Faith: Corruption and the Failure of Law Reform. In: pp. 278-292. C. Manning and P. van Diermen (eds.). Indonesia in Transition: Social Aspects of Reformasi and Crisis. Institute of Southeast Asian Studies, Singapore.
Logical Flaws in Regional Autonomy. 2000. Jakarta Post. 2 May.
Lowry, R. 1996. The Armed Forces of Indonesia. St. Leonards, Australia: Allen & Unwin.
Lowry, R. 1999. Indonesian Armed Forces (Tentara Nasional Indonesia–TNI). Research Paper No. 23, 1998-99. Parliament of Australia, Canberra.
Lynch, O. and K. Talbott. 1995. Balancing Acts. World Resources Institute: Washington, DC.
A Matter of Law . . . and, Of Course, Order. 2000. The Economist. 8 July.
McCarthy, J.F. 2000. 'Wild Logging': The rise and Fall of Logging Networks and Biodiversity Conservation Projects on Sumatra's Rainforest Frontier. Center For International Forestry Research, Jakarta.
McCullouch, L. 2000. Business as usual: Until Gus Dur can bring military business activities under control, they won't go 'back to the barracks.' Inside Indonesia 63 (July-September). Retrieved from the World Wide Web: http://www.insideindonesia.org/edit63/mccullochl.html.
Ministry of Forestry and Estate Crops. 2000. Analysis and Discussion Paper by the Director General for Protection and Conservation of Nature (translation from original by the author). National Working Meeting of the Ministry of Forestry and Estate Crops, June 26-29, 2000, Jakarta.
Ministry of Forestry and Plantations. 2000. Rencana Stratejik 2001-2005 [Strategic Plan 2001-2005]. Jakarta.
Mittermeier, R. A., N. Myers, and C.G. Mittermeier. 1999. Hotspots. Cemex: Mexico City.

National Development Planning Agency (BAPpENAS). 1993. Biodiversity Action Plan for Indonesia. Jakarta.

Peluso, N.L. 1993. Coercing Conservation: The Politics of State Resource Control. Global Environmental Change. 4 (2):19-217.

Potter, L. and J. Lee. 1998. Tree Planting in Indonesia: Trends, Impacts, and Directions (consultancy report for the Center for International Forestry Research [CIFOR]). Adelaide, Australia.

Reform of the Legal System Indonesia's Top Challenge: Minister. 2000. Agence France-Presse. 9 March.

Saunders, J. 2000. Indonesian Forces are Part of the Problem in the Moluccas. International Herald Tribune. 4 July.

Schwarz, A. 1994. A Nation in Waiting: Indonesia in the 1990s. Allen & Unwin, St. Leonards, Australia.

Scott, M. 1998. Indonesia Reborn? New York Review of Books. 13 August.

Sunderlin, W.D. and I.A.P. Resosudarmo. 1996. Rates and Causes of Deforestation in Indonesia: Towards a Resolution of the Ambiguities. Center for International Forestry Research, Bogor, Indonesia.

Talbott, K. 1998. Logging in Cambodia: Politics and Plunder. In: F. Brown and D. Timberman. Cambodia: A Way Forward. The Asia Society, New York.

UNHCR (United Nations Office of the High Commissioner for Human Rights). 2000. Report of the International Commission of Inquiry on East Timor to the Secretary-General. UNHCR, Geneva.

Workshop Questions Indonesia's Autonomy Laws . . . and Expert Urges 'Some Form of Federalism.' 2000. Jakarta Post. 18 July.

The World Bank. 1998. Indonesia in Crisis: A Macroeconomic Update. Washington, DC.

The World Bank. 2000a. The Challenges of World Bank Involvement in Forests: An Evaluation of Indonesia's Forests and World Bank Assistance. Washington, DC.

The World Bank. 2000b. Deforestation in Indonesia: A Preliminary View of the Situation in 1999 (Draft Report). Jakarta.

The World Bank. unpublished memorandum. World Bank Involvement in Sector Adjustment for Forests in Indonesia: The Issues. Jakarta.

Vincent, J. M Gillis. 1998. Deforestation and Forest Land Use: A Comment. The World Bank Research Observer, no. 13:133-140.

Zerner, C. 1992. Indigenous Forest-Dwelling Communities in Indonesia's Outer Islands: Livelihood, Rights, and Environmental Management Institutions in the Era of Industrial Forest Exploitation (consultancy report prepared for the World Bank Indonesia Forestry Sector Policy Review). Resource Planning Corporation, Washington, DC.

Legal Mechanisms for Addressing Wartime Damage to Tropical Forests

Jay E. Austin
Carl E. Bruch

SUMMARY. The tactics of war have profound impacts on tropical forest ecosystems, and modern weapons technologies have greatly increased their destructive potential. Some legal protection is afforded by customary international law, and the international community responded to the Vietnam War by adopting, *inter alia*, the 1977 Additional Protocol I to the Geneva Conventions and the 1976 Environmental Modification Convention, which prohibit "long-term" (or "long-lasting"), "widespread," and "severe" environmental damage. Nevertheless, many regard these and other existing conventions as inadequate, especially as applied to internal conflicts. More recently, the International Union for the Conservation of Nature-World Conservation Union (IUCN) put forth a Draft Convention on the Prohibition of Hostile Military Activities in Internationally Protected Areas. This paper analyzes the IUCN Draft Convention–particularly issues of prior designation of protected areas, waiver of protection, and monitoring and enforcement–and compares it to other relevant area-based treaties, such as the 1954 Hague Convention for the Protection of Cultural Property and the 1972 World Heritage Conven-

Jay E. Austin and Carl E. Bruch are Senior Attorneys, Environmental Law Institute, 1616 P Street NW, Washington, DC 20036.

The authors wish to dedicate this article to the memory of Bernard Nietschmann.

Financial support for this article was provided by the John D. and Catherine T. MacArthur Foundation.

[Haworth co-indexing entry note]: "Legal Mechanisms for Addressing Wartime Damage to Tropical Forests." Austin, Jay E., and Carl E. Bruch. Co-published simultaneously in *Journal of Sustainable Forestry* (Food Product Press, an imprint of The Haworth Press, Inc.) Vol. 16, No. 3/4, 2003, pp. 167-199; and: *War and Tropical Forests: Conservation in Areas of Armed Conflict* (ed: Steven V. Price) Food Products Press, an imprint of The Haworth Press, Inc., 2003, pp. 167-199. Single or multiple copies of this article are available for a fee from The Haworth Document Delivery Service [1-800-HAWORTH, 9:00 a.m. - 5:00 p.m. (EST). E-mail address: getinfo@haworthpressinc.com].

tion. The paper also highlights other recent legal developments that may help prevent, remediate, or punish wartime damage to tropical forests, such as the nascent International Criminal Court; liability mechanisms for providing compensation for wartime damage; environmental guidelines for military, peacekeeping and humanitarian operations; import bans and trade embargoes; and a proposed no-fault remediation fund.

KEYWORDS. International environmental law, international law of war, warfare, military activities, protected areas, tropical forests

INTRODUCTION

War can devastate tropical forests. Combatants deliberately target forests to deprive their opponents of potential troop cover. Rebel and government forces pillage their country's natural heritage (especially valuable resources such as ivory, teak, and diamonds) to finance military activities. Refugees fleeing armed conflict frequently settle in or near sensitive forest ecosystems, cutting trees for cooking and heating fuel. Furthermore, both during and after conflict, individuals and companies take advantage of the disorder and confusion to poach wildlife, log forests, and extract minerals without any of the normal legal protections or community oversight.

This paper explores legal mechanisms for preventing, remediating, and punishing wartime damage to tropical forests. Section II describes the factual background and identifies several different types of impacts that war and related activities can have on tropical forests. Section III sketches the existing international legal framework for addressing wartime damage to forests and other ecosystems, and discusses its limitations. Section IV presents emerging and proposed legal norms and institutions that could prevent, remediate, or punish wrongful damage to tropical forests, and highlights a proposed international convention, drafted by the International Union for the Conservation of Nature–World Conservation Union (IUCN), that seeks to protect particularly important ecosystems by designating them off-limits to military activities. We analyze this area-based proposal in depth, and place it in the context of other emerging and proposed legal mechanisms aimed at lessening the environmental impacts of war through criminal liability, civil liability, military guidelines, import bans, and emergency remediation.

FACTUAL BACKGROUND

Environmental impacts of armed conflict generally fall into four categories: the effects of deliberately or indiscriminately targeting the environment; the plundering of natural resources to support or finance military operations; the impacts caused by refugees and displaced persons as they seek food, shelter, and heat; and the broader consequences of an armed and lawless society. Each of these categories lends itself to different kinds of legal and institutional responses. In general, deliberate and indiscriminate actions that target the environment can be deterred and punished by state responsibility and liability and individual criminal liability, and international treaties and institutions play a significant role here. Similarly, when armed conflict is being supported by exploitation of natural resources, international embargoes and commodity import bans may help hasten peace. However, at the international level, there are significantly fewer legal options when it comes to addressing the impacts of refugees and the general lawlessness associated with armed conflict and its aftermath.

Targeting the Environment

Tropical forests and other complex ecosystems offer troops shelter, food, water, fuel, and medicine. As a result, opposing combatants frequently target the natural environment, burning and defoliating forests, poisoning water supplies, and even attempting to change weather patterns. In addition, certain weapons, such as anti-personnel landmines and cluster bombs, affect a population's movement and land-use patterns and have long-lasting consequences for non-combatants–civilians, animals, and the environment–for many years after their military purpose has been served.

While destruction of the landscape is as old as war itself, the Vietnam War highlighted the increasingly devastating environmental effects of modern military technology, and became a watershed event in spurring action to constrain environmental consequences of war. To hamper Viet Cong troop movements and provide protection for river patrols, U.S. forces engaged in a massive defoliation campaign, spraying herbicides over 10% of South Vietnam and scraping the topsoil from another 2%. They also attempted to change weather patterns via cloud seeding over North Vietnam and elsewhere in Indochina (Westing 1976; National Academy of Sciences 1974; Westing 1977; U.S. Senate Committee on Foreign Relations 1972).

Thirty years later, the *New York Times*, 16 May 1999, reported that large tracts of former Vietnamese forests remain treeless. Moreover, as reported in the *Washington Post*, 18 April 2000, the public health implications of environmental warfare in Vietnam–primarily birth defects, diseases, and premature death associated with exposure to Agent Orange–also have become apparent, and have been the subject of much study (Schecter et al. 1995; Savitz et al. 1993; Dreyfuss 2000). However, concrete data on the health impacts remains elusive. While Peter Waldman has reported that Vietnamese scientists believe that since the mid-1960s, as many as 500,000 people may have been born with dioxin-related deformities, but Alastair Hay has found the epidemiological data to be less clear (Waldman 1997; Hay 2000).

Notwithstanding international condemnation of military defoliants, Central American internal conflicts of the 1980s saw further use of defoliation and other tactics harmful to the environment (Weinberg 1997; Weinberg 1991; Hall and Faber 1989). For example, in El Salvador, some areas saw damage to up to 60% of the standing trees, heavy-metal contamination of the ground, clay soils hardened by incendiary weapons, softer soils exposed to erosion, and landmines continuing to harm people and wildlife (Navarro 1998).

Indeed, small, cheaply produced landmines pose a persistent and indiscriminate threat to civilians and wildlife throughout the world. Although often laid to protect bases or other strategic areas, they are rarely mapped or marked, and maim and kill civilians and wildlife long after their military purpose has passed. Many of the most heavily mined countries–such as Cambodia, Colombia, and Mozambique–also host rich biological diversity. In Colombia, guerrilla groups operate out of remote forest and jungle regions; some have their bases in designated national protected areas, and the numerous mines planted by army and guerrilla troops have devastating impacts on forest wildlife and people (Rodriguez 1998). While reliable data on the impacts of mines on wildlife is scarce, Kevin Steward has estimated that globally, mines have killed more than 1.6 million animals in 39 countries (Whaley 2000).

Financing Wars with Natural Resources

Combatants also have been known to plunder natural resources–including tropical forests and the minerals and wildlife within them–to pay their troops and purchase weapons. For example, Kirk Talbott reported that Cambodia's timber stocks have been and continue to be exploited to finance the political and military rivalries, with the Khmer Rouge at one point receiving US$10-20 million a month from logging (Talbott 1998). In the late 1980s, the Angolan rebel group UNITA (National Union for the Total Independence of Angola) paid for South African military support with teak and ivory (Sayagues 1999). The *Washington Post*, 16 March 2000, reported that UNITA has earned more than US$5 billion from diamond mines they control; while *The Economist*, 15 January 2000 observed that the Angolan government has exploited petroleum reserves–estimated at US$3.5 billion a year–to finance its military campaign. Elsewhere, the *New York Times*, 8 August 1999, cited a U.S. administration official's estimate that the civil war in Sierra Leone, which killed and maimed thousands, has been prolonged by at least 18 months due to the rebels' ability to trade their diamonds for arms. Timber and valuable minerals financed Charles Taylor's coup in Liberia (Boutwell and Klare 2000), and they continue to motivate and support the various parties in the war in the Democratic Republic of the Congo (Report of the Panel of Experts 2001).

Over the last decade, Colombian forests have been cleared for cultivation of marijuana, coca, and heroin poppies, which provide a significant source of funding for various guerrilla groups. Approximately 100,000 ha of land are currently cleared for drug crops, including Andean cloud forests, which recover slowly (Rodriguez 1998). This deforestation in turn threatens the water sources that feed 70% of Colombia's rivers (Weiskopf 1999). Additionally, the intensive use of fertilizers, pesticides, and other chemicals associated with drug cultivation and processing affects the fragile soils, waters, and biological web of life in the rainforests, particularly as large commercial drug plantations move further into the Amazon and Orinoco rainforests (Rodriguez 1998; Weinberg 1999; "Colombian Army Cites Ecosystem Damage" 2001).

Moreover, the use of natural resources to finance militaries invites attacks on those same resources. For years, Colombian rebels have been conducting economic warfare by detonating petroleum pipelines. This has spilled at least two million barrels of crude oil into rivers, contaminating drinking and irrigation water, killing fish and other wildlife, contributing to forest fires and air pollution, sterilizing soil, and harming riverside communities ("Colombia Urges" 1998; Bruch 1998a). By 1998, the affected areas were estimated to include 2,600 km of watercourses, 1,600 ha of wetlands, and 6,000 ha of agricultural land (Rodriguez 1998). The Colombian government's estimated damages include US$26 million in lost crude oil, another US$26 million in environmental clean-up costs, and US$1.5 billion in lost oil revenues ([Colombian] Ministerio del Medio Ambiente 1998). The terrestrial and riverine impacts have extended beyond national borders to Venezuela. Similarly, in Sudan rebels have begun to bomb an oil pipeline that will provide the Sudanese government with significant export revenues that some anticipate will be used to buy more weapons. The *New York Times*, 16 February 2000, reported that the Canadian government, while investigating a Canadian oil company operating in Sudan, found that the existence of the new oil industry there is aggravating the civil war.

At the same time, the Colombian government, with the active support of the U.S. government, has undertaken a campaign to eradicate drug crops through aerial spraying. Diana Jean Schemo of the *New York Times*, 20 June 1998, reported that the Colombian government was going to test application of tebuthiuron, a powerful broad-spectrum herbicide. However, Dow Chemical warned that its product should not be used for that purpose because application "can be very risky in situations where terrain has slopes, rainfall is significant, desirable plants are nearby and application is made under less than ideal circumstances." Tim Golden also reported in the *New York Times*, 6 July 2000, on the debate over the use of fungus to attack coca fields in Colombia.

Aside from depriving a country of capital that is desperately needed for development or social programs, financing wars with natural resources prolongs the misery of war and often wreaks greater environmental harm, as constraints and mitigation requirements that may be placed on resource extraction during peacetime are ignored in the urgency of conflict. This emphasis of short-term gains over long-term sustainability drains national resources and makes it more difficult to return to peaceful life after the conflict.

Refugees and Displaced Persons

Refugees and displaced persons pose a particularly challenging threat to the environment. Obviously, there is a humanitarian need to supply the basics–shelter, food, water, and heat–to people who no longer have access to their normal resources or infrastructure. When faced with hundreds of thousands of refugees, international humanitarian agencies sometimes direct them to settle on undeveloped land and to make use of available natural resources. Unfortunately, these large populations require equally large amounts of firewood, food, and water, and have minimal infrastructure available for handling their needs, including sewage and waste disposal.

Thus, following the Rwandan genocide, the United Nations High Commissioner for Refugees (UNHCR) permitted refugees to cut firewood in the Virunga National Park in the Democratic Republic of the Congo, a United Nations (UN)-listed World Heritage site. They cut an estimated 47 million cubic yards of firewood, which devastated the habitat of endangered mountain gorillas (Schmidt 1999). Further, *Down to Earth*, 15 May 1999, reported that refugees have settled in two-thirds of Akagera National Park, the largest park in Rwanda, and refugees in both Virunga and Akagera National Parks have poached gorillas, hippos, and other wildlife. The story is similar with respect to people fleeing conflict in other tropical areas, including the Upper Guinea forest, which is impacted by refugees from Liberia and Sierra Leone. In Colombia, the ongoing conflict is estimated to have displaced 270,000 to 310,000 persons, with many moving to frontier areas, where they clear virgin forests to develop agricultural land (Rodriguez 1998).

The dramatic impacts of refugees in or near tropical forests are aggravated by the resulting commoditization of natural resources. Refugees and local residents often engage in profiteering, cutting more firewood and killing more wildlife than they need in order to sell it to a market for these materials that has suddenly swelled. For example, in the late 1980s, UNHCR found that in Honduras it was purchasing from local markets so much firewood for Nicaraguan refugees that the forests became significantly impacted (Jacobsen 1994).

Environmental Impacts of Lawlessness

During armed conflict, the general culture of lawlessness can have dire impacts on the environment (Navarro 1998). Police and park ranger resources are commandeered to support combat efforts, law enforcement officials become more concerned with simply maintaining civil order than with stopping illegal poaching or logging, and a heavily armed population readily turns its weapons on wildlife. Thus, Ian Fisher reported in the *New York Times*, 28 July 1999, that poaching of gorillas and elephants is thriving in the wake of Congo's war now that poachers are armed with automatic weapons. For five years now, there have been no tourists, whose presence had "helped keep poachers away, in part because so many people were in the forest. As important, their money helped build schools, roads and water projects that made people in the poor and densely populated area around the park feel they had a stake in its survival."

In Cambodia, illegal and uncontrolled commercial logging was rampant during the government's prolonged struggle against the Khmer Rouge (Talbott 1998; Peters 1999). Another legacy of this armed conflict has been the profusion of landmines, machine guns, and artillery shells, which are now used to trap and kill tigers and other wildlife, coupled with the psychological and sociological effects of prolonged warfare. Thus, one researcher in Cambodia observed that "It's not just the mines and artillery shells. It's thirty years of everybody having a gun and mowing down everything they see in the countryside" (Whaley 2000, 27). Likewise, on the cultural front, Seth Mydans of the *New York Times*, 1 April 1999, reports that Cambodian temples have been looted, with tons of stone sculptures and inscriptions carted away, both during the conflict and after its end.

Conversely, the harshness of war occasionally can benefit the environment (McNeely 2000; Daltry and Momberg 2000). During the Nicaraguan civil war of the 1980s, timber felling was sharply reduced; conversion of forests to agricultural land slowed and stopped; animal trafficking largely halted; and fishing harvests fell as fishermen who feared naval mines stayed ashore, allowing depleted stocks to restore themselves (Girot and Nietschmann 1992). Many of the remaining intact ecosystems in Central America continue to be threatened with conversion by agrarian peasants, with the exception of those that were mined during the civil wars. Similarly, probably the most biologically diverse area on the Korean Peninsula is the heavily mined area around the Demilitarized Zone, a four-kilometer-wide no-man's-land where the 1953 Armistice Treaty prohibits development and hunting (Kim 1997; Westing 1998-99). However, while land littered with mines and unexploded ordnance may have a brief respite, from a humanitarian and environmental perspective the overall picture remains grim. In addition to ongoing human and wildlife casualties, people displaced from mined farmland migrate to forests and clear trees for agriculture and kill wildlife for subsistence purposes (Whaley 2000).

THE EXISTING LEGAL FRAMEWORK

The long litany of wartime impacts to forests and wildlife raises the question of what international law can do to address them, and what enforcement capability it can lend to situations where the rule of domestic law has broken down. Relevant bodies of law include the international law of war, both generally and as it relates specifically to environmental impacts, as well as the flourishing body of international environmental law. However, while the law of war addresses some of the situations described, it can be vague, incomplete, and difficult to enforce, especially in internal conflicts; and peacetime environmental law, while promising, will require further development and some creative thinking before it can be applied to wartime settings.

The International Law of War

Historically, the law governing environmental consequences of war has been rooted in international humanitarian law, in particular the law of war. The law of war provides guidance through customary norms of necessity, proportionality, discrimination, and humanity, as well as through a series of treaties, the Geneva Conventions and Hague Conventions, that incorporate these norms and govern the behavior of combatants and the use of weapons (Austin and Bruch 2000). Even without expressly mentioning the environment, this body of law affords some protection to natural ecosystems, in the same way as it does to any other area or target that is not essential for military objectives. It has been criticized for the difficulty of interpreting and applying its terms (Schmitt 2000), as well as for the arbitrariness with which it has been enforced (Falk 2000). Yet military and civilian commentators alike agree that these principles constitute a useful starting point for wartime environmental protection, at least as it relates to "collateral" (incidental or unintentional) damage and to the largely anthropocentric goal of preserving living conditions and resources essential for the survival of the civilian population.

The Vietnam War, with its systematic targeting of forests, gave rise to new concerns about ecosystem damage and spurred a corresponding rethinking of the law of war. The international community developed two new law-of-war treaties that were the first explicitly addressing the environment: the 1976 Convention on the Prohibition of Military or Any Other Hostile Use of Environmental Modification Techniques (ENMOD), and the 1977 Additional Protocol I to the Geneva Conventions (Protocol I). The more important of the two, Protocol I, not only codified and expanded customary law-of-war principles, but also expressly prohibited "methods or means of warfare which are intended, or may be expected, to cause widespread, long-term, and severe damage to the natural environment" (Protocol I, Articles 35(3) & 55). However, it did not define the key terms "widespread," "long-term," and "severe," though one quasi-official comment suggests that "long-term" damage would properly be measured in decades. While several major powers such as the United States and United Kingdom have not ratified Protocol I, it is recognized to reflect many norms of customary international law and the U.S., U.K., and other countries in practice have incorporated many Protocol I provisions into their military manuals.

ENMOD, which uses similar language, has been interpreted to define "widespread" damage as meaning on a scale of several hundred square kilometers, "long-lasting" as a period of months, and "severe" as "involving serious or significant disruption or harm to human life, natural and economic resources, or other assets." Unfortunately, ENMOD's scope is narrow, addressing only instances where the environment itself is deliberately used as a weapon, such as bombing dams to release a wall of water or attempting to activate earthquake faults or dormant volcanoes. Thus, as a result of both treaties' limitations, commentators still disagree whether the actual conduct of the Vietnam War would have been proscribed by the very legal norms that were drafted in response to it (Austin and Bruch 2000).

In short, both Protocol I and ENMOD were created to extend wartime protections to the natural environment. In contrast to customary law-of-war principles, which generally entail a case-by-case balancing of the collateral harm caused by an attack against the military advantage gained, both treaties attempted to set an objective–though ill-defined–threshold for excessive environmental harm. If it is established that "widespread, long-term and severe" environmental damage has occurred, warring parties arguably could be held responsible and liable regardless of where their attack took place, whether it had a military objective, or whether the environmental damage was intentional. In this sense, these treaty provisions represent a rare instance of ecological values trumping the oft-asserted exigencies of armed conflict–what Richard Falk calls "the sacred cow of military necessity" (Falk 2000).

However, this drafting achievement may have come at the expense of enforceability, as those faced with interpreting and applying the treaty language likely will be inclined to define such a high threshold of environmental damage that it will render the provisions inapplicable in all but the most egregious cases. Alternatively, commentators and tribunals may be inclined to re-import customary law principles in order to preserve some kind of military necessity defense. Indeed, this appears to have happened in the negotiations leading to creation of the International Criminal Court, where identical treaty language prohibiting "widespread, long-term, and severe" environmental damage was weakened by stipulating that such damage must be "clearly excessive" in relation to the military advantage gained (Rome Statute, Article 8(2)(b)(iv); Austin and Bruch 1999). The 1980 Convention on Certain Conventional Weapons (CCW) and its Protocol III provide another such example. The preamble of the CCW Convention echoes the prohibition against "widespread, long-term and severe" environmental damage, and Article 2(4) of Protocol III prohibits attacks on "forests or other kinds of plant cover" using incendiary weapons. However, the protections are waived "when such natural elements are used to cover, conceal or camouflage combatants or other military objectives, or are themselves military objectives." That armies are prohibited from using incendiaries to attack forests, but have recourse to a military necessity exception that in practice threatens to swallow the prohibition.

Protocol I, ENMOD, and other existing law-of-war treaties that provide some norms and institutions for addressing the environmental consequences of armed conflict, apply only to international conflicts between two or more states. The norms applicable to internal armed conflicts are much more circumscribed. For example, Additional Protocol II to the Geneva Conventions, which deals with conduct of internal conflicts, does not contain an express prohibition against environmental destruction, and the Rome Statute of the International Criminal Court similarly failed to proscribe environmental war crimes in non-international settings. (Rome Statute, Article 8(2)(e); Austin and Bruch 1999). However, most conflicts since World War II have been of this latter variety: 97 of the 103 armed conflicts between 1989 and 1997 were civil wars or other internal strife (Renner 1999).

International Environmental Law

In view of uncertainty about the scope and effectiveness of law-of-war treaties for protecting the environment during armed conflict, it is helpful also to examine norms, rules, and enforcement mechanisms derived from other international legal regimes. The environmental protection systems that have evolved under peacetime conditions generally are more nuanced and tailored to their subject matter than the sporadic attempts to incorporate environmental concerns into the law of war. While they generally do not address the extreme circumstances encountered in full-scale armed conflict, many peacetime environmental treaties have state-of-the-art provisions for deterring and remedying the same kinds of impacts–including deforestation and species loss–commonly found during wartime. Thus, at least where it is possible to consider these impacts in isolation, peacetime environmental laws may prove useful for wartime protection of tropical forests, either through their direct application or through analogizing and adapting them into the law-of-war regime.

One obvious place to begin is with treaties whose primary aim is the delineation and protection of natural areas, in particular the 1972 United Nations Educational, Scientific and Cultural Organization (UNESCO) Convention Concerning the Protection of the World Cultural and Natural Heritage (World Heritage Convention). With its rationale based on preventing "destruction not only by the traditional causes of decay, but also by changing social and economic conditions which aggravate the situation with even more formidable phenomena of damage and destruction" (Preamble), this Convention and its World Heritage List are plausible mechanisms for preventing and addressing wartime damage. Indeed, one of the stated criteria for inclusion on the List is "the outbreak or threat of an armed conflict" (Article 11(4))–although commentators have noted that there are no specific provisions for actual wartime implementation (Burhenne 1997; Tarasofsky 2000). At a minimum, the Convention requires states "not to take any deliberate measures which might damage directly or indirectly the cultural and natural heritage (Article 6(3)), and to adopt "education and information programmes . . . to strengthen appreciation and respect by their peoples of the cultural and natural heritage" (Article 27(1)). The latter provision arguably could include training military forces to observe the international protections.

However, the World Heritage Convention is not only silent on how it could be applied to wartime situations, it has few enforcement provisions generally, and is chiefly a vehicle for financial and technical assistance. Further, its express goal of "fully respecting the sovereignty of the states on whose territory the cultural and natural heritage . . . is situated" (Article 6(1)) suggests that it likely would have limited reach over military actions taken in self-defense, or over purely internal conflicts. As a practical matter, as noted above, many of the ecologically sensitive areas threatened by armed conflict in Africa, Central America, and elsewhere have already been named World Heritage Sites–but to little effect in the face of troop movements or large refugee populations. Absent a more focused attempt to extend the Convention to cover these issues or to instill its values into military operations, it appears inadequate to the task of regulating wartime environmental damage.

Another question is whether global environmental treaties, such as the Convention on Biological Diversity, the Framework Convention on Climate Change, or a potential Forestry Convention, remain in effect during armed conflict. If it could be proven, for example, that wartime destruction of forests had led to loss of significant species habitat or carbon sinks, those treaties could serve as an additional basis for holding the party that caused the deforestation responsible and liable under international law. Silja Vöneky has argued that, if a peacetime environmental treaty is silent as to its applicability, whether it will be applicable and binding during wartime depends on the character of the treaty (Vöneky 2000, 193). According to Vöneky, peacetime environmental treaties should remain applicable during wartime to the extent that they purport to protect "the interests of the state community as a whole." Applying this standard, major area-based treaties, possibly including the World Heritage Convention, could serve as a source of norms governing military activities within their respective areas of jurisdiction. In addition, treaties that regulate the global commons, such as the stratosphere or the biosphere, also could remain binding on belligerent states. Vöneky's approach is promising, especially since many details of implementing both the Biodiversity Convention and the Climate Change Convention have yet to be worked out. As for a future Forestry Convention, the simplest approach of course would be to draft the text of any such treaty to ensure that it expressly applies to wartime situations.

EMERGING AND PROPOSED NORMS AND INSTITUTIONS

Whether norms and definitions of unacceptable wartime damage to tropical forests and other ecosystems come from the law of war or from international environmental law, the real issue is how they can be translated into practice. Toward this end, a number of recent proposals have focused not only on clarifying and refining the existing sources of international law, but also on establishing concrete procedures and institutions for implementing and enforcing their goals. While each of these proposals seeks in some degree to prevent, remediate, and punish wartime environmental damage, they vary according to which of these objectives is dominant, the kinds of legal mechanisms they employ, and their suitability for addressing the different types of wartime threats to tropical forests. On a loose analogy to mechanisms used in peacetime environmental law, they can be classified as area-based protection, criminal liability, civil liability, operating guidelines, import bans and trade embargoes, and emergency remediation. The analysis in this section focuses on the first category, and then summarizes recent developments in the remaining five categories.

Area-Based Protection

Rather than attempting to define and to proscribe a specific level of unacceptable environmental damage, another strategy, directly relevant to forest protection, would be to declare ecologically sensitive areas to be off-limits to military activity altogether. As seen above, that is the theoretical promise of applying the World Heritage Convention to wartime situations; but it also could be accomplished through extending the law of war, by analogy to existing treaties that aim to protect hospitals, churches, museums, and other specific sites on humanitarian and cultural grounds. This approach is embodied in a recent draft convention put forth by the International Union for the Conservation of Nature-World Conservation Union (IUCN).

IUCN's Draft Convention on the Prohibition of Hostile Military Activities in Protected Areas is the first focused attempt at an area-based system of wartime environmental protection. (Burhenne 1997; Tarasofsky 2000). As such, it descends from a long line of law-of-war treaties that restrict entire areas or categories of targets, such as the 1954 Hague Convention for the Protection of Cultural Property in the Event of Armed Conflict. Like that earlier convention, the IUCN Draft Convention would require designating, ideally in advance of conflict, special protected areas in which no hostile military activities would be permitted, subject to certain clearly defined exceptions. Such legal regimes are largely aimed at deterring harmful activity, but they also have the advantage of establishing bright-line rules that in theory make it easier to discover, prove, and sanction violations when they occur. Toward this end, the Draft Convention also provides for compliance monitoring by UN-supervised expert missions, compulsory arbitration of inter-party disputes, and the possibility of referring disputes to the International Court of Justice.

Under the IUCN Draft Convention, the UN Security Council would be required to include in its resolutions dealing with armed conflict a list of "natural or cultural area[s] of outstanding international significance" in which military activities would be prohibited (IUCN Draft Convention, Article 1(a)). Site selection would be done ad hoc as conflicts are anticipated, but weight could be given to existing designations of international protected areas, such as UNESCO World Heritage Sites or

Biosphere Reserves. Once designated by the Security Council, the sites would become "non-target" areas for the duration of the conflict (Article 2). However, this protected status is forfeited if the nation where the area is located either maintains a military installation within the area or decides to use it to carry out military activities (Article 3). The forfeiture provision in effect enables the host nation to waive protected status, a result that arguably is consistent with a nation's sovereign right to elevate its self-defense over environmental considerations (Tarasofsky 2000). Similarly, once protected status is waived, an attacking force presumably then would be free to employ the justifications of military necessity and proportionality in deciding whether and how to act against military objectives located in the area.

There are a number of questions raised by the Draft Convention in its current form. First, relying on the UN Security Council to apply and implement its provisions may be problematic. Wolfgang Burhenne and Richard Tarasofsky argue that the Security Council's authority to issue environmental prohibitions is inherent in its power to respond to armed conflict under Chapter VII of the UN Charter. They analogize to the UN "safe areas" declared on humanitarian grounds during the Bosnia conflict (Burhenne 1997; Tarasofsky 2000). However, Tarasofsky notes the interesting legal issue of whether any treaty other than the UN Charter itself can "require" the Security Council to take action. He observes that the Draft Convention technically may only be able to "request" that the Security Council make the protected area designations through whatever method it deems appropriate.

Equally problematic is the matter of monitoring and enforcing protected areas once they are designated. The horrifying experience with "safe areas" in Srebrenica and Tuzla, Bosnia, as well as the ongoing problems in World Heritage Sites, suggest that UN protection on paper–or even on the ground–is often woefully inadequate to the realities of modern warfare. The Draft Convention does provide for UN-sponsored expert monitoring missions, whose members would be granted the same status as UN personnel, and requires the sending body "to take necessary actions to ensure effective implementation" of the Convention (IUCN Draft Convention, Article 4). However, once violations are discovered, it is far from clear whether the Security Council would have the authority–much less the political will–to deploy peacekeeping forces solely to prevent or to halt wartime environmental damage. The Draft Convention's dispute resolution and arbitration mechanism provides another potential remedy for violations (Article 7). But it is one that would seem either to rely on nations to declare in advance their intentions to attack an area whose protected status is in dispute, or to be limited to remediation of environmental damage after it has already occurred.

A final problem is that, here again, the IUCN Draft Convention is silent as to whether it only applies to international conflicts between two or more states, or also applies to civil wars or other internal conflicts within a state. As with ENMOD, Protocol I, and other law-of-war treaties, restricting its application to international conflict would severely limit its usefulness. Tarasofsky takes an optimistic view of this drafting omission, claiming that "the Draft Convention appears to cover both cases" (Tarasofsky 2000, 572). But it seems equally plausible that future drafters or interpreters of the Convention may conclude that it should not apply to internal conflict, as happened to the environmental damage provision of the Rome Statute of the International Criminal Court. Thus, the scope and jurisdiction of the Draft Convention should be expressly broadened in any future drafts, rather than leaving it open to interpretation.

Indeed, by its terms, the IUCN Draft Convention is a work in progress, intended to stimulate further discussion, thought, and revisions before advancing to a full-fledged negotiating process. Hopefully, the impetus for such a process will come from IUCN's ongoing efforts to promote the Draft Convention, as well as the recent publication of its full text and an accompanying analysis (Tarasofsky 2000). Moreover, it may prove possible to build upon the IUCN Draft Convention by taking account of another recent effort–the Second Protocol to the 1954 Hague Convention for the Protection of Cultural Property. Opened for signature in 1999 but not yet entered into force, the Second Protocol tries to resolve some of the same enforcement issues surrounding wartime protection of cultural property that have been raised for protection of the environment. As such, and given the similarities in both treaties' area-based approach, it is a rich source of potential improvements and refinements to IUCN's environmental initiative.

The original 1954 Cultural Property Convention was drafted partly in response to Nazi atrocities, and created a regime that deems all cultural property–religious or secular, historical or artistic, movable or immovable–to be protected from damage and plunder during armed conflict. It requires states to safeguard cultural property on their territory from the "foreseeable effects of an armed conflict" (Article 3), and to refrain from any acts of hostility against cultural property on their own or other states' territory or uses that "are likely to expose it to destruction or damage in the event of armed conflict" (Article 4(1)). These obligations may be waived only "where military necessity imperatively requires such a waiver" (Article 4(2)). Further, a state may place a limited number of cultural sites under "special protection," which enrolls them in an International Register and immunizes them from acts of hostility or military use (Articles 8, 9). Immunity may be withdrawn only upon a violation by the side claiming immunity, or "in exceptional cases of unavoidable military necessity" (Article 11). These provisions may be enforced between states through a special conciliation procedure (Article 22), and against individuals through criminal sanctions for persons who breach the Convention or order that it be breached (Article 28).

In practice, the Cultural Property Convention has been little-used, and has yet to attain its goal of popularizing a "Blue Shield" emblem and ethic that would command the same level of respect for cultural property as the Red Cross does for humanitarian concerns (Burhenne 1997; Tarasofsky 2000). Cognizant of this gap between theory and practice, the 1999 Second Protocol to the Convention is a detailed instrument that will supplement and almost completely rewrite the original text, providing new definitions and procedures. It outlines clear standards for making a determination of military necessity, requiring that the property be a legitimate military objective, that there is "no feasible alternative available to obtain a similar military advantage," and that an advance warning of the attack be given when circumstances permit (Article 6). Even when an attack is permitted, the attacking party must "take all feasible precautions . . . with a view to avoiding, and in any event to minimizing, incidental damage" to protected cultural property (Article 7). The Second Protocol also establishes an additional level of "enhanced protection" for cultural property "of the greatest importance for humanity," and grants near-absolute immunity to such property unless it is being actively used for military purposes (Articles 10, 12).

Equally important, the Second Protocol greatly expands upon the enforcement provisions of the original Cultural Property Convention and establishes an enforcement scheme comparable to the rest of international humanitarian law. It not only provides for individual criminal responsibility for serious, intentional violations (Article 15), it obligates states to adopt domestic legislation that defines these violations and makes them punishable, to prosecute or extradite violators, and to provide mutual assistance in investigations and prosecutions (Articles 15-18). States also are obligated to punish other, less serious, violations of the Convention and the Protocol through appropriate legislative, administrative or disciplinary measures (Article 21).

Finally, the Second Protocol expressly applies, in its entirety, to "armed conflicts not of an international character, occurring within the territory of one of the Parties" (Article 22). This expands upon the comparable section of the original Convention, which tentatively extended to internal conflicts its general provisions on respect for cultural property, but not the provisions on specially protected areas (Cultural Property Convention, Article 19). Thus, while the Second Protocol makes careful assurances about preserving state sovereignty, and distinguishes between full-scale internal conflict and merely "internal disturbances and tensions, such as riots, isolated and sporadic acts of violence and other acts of a similar nature," it nevertheless clearly intends its protections and enforcement provisions to apply to all sides in civil wars and guerrilla wars. In this regard, it remedies the chief deficiency of most existing and proposed law-of-war treaties that address the environment, including the IUCN Draft Convention.

In short, there is much in the Second Protocol to the Cultural Property Convention that could serve as a model for fleshing out the IUCN Draft Convention or a similar area-based scheme for wartime nature protection. Moreover, the Second Protocol's stringent test for military necessity–couched in the same language of "alternatives," "precautions," and "minimization" that is used in peacetime environmental impact assessments–could shed some light on how to apply the vague norms of customary international law to specific targeting decisions and specific cases of environmental damage. Its enforcement regime spells out all the elements necessary for criminal prosecution of intentional violations, which helps ensure that the only possible obstacles to such prosecutions will be practical and political, rather than legal.

Criminal Liability and Punishment

As just seen, even area-based environmental protection, which largely aims at preventing damage, can encompass criminal enforcement for certain offenses once they occur. The same is true for customary international law and the relevant law-of-war treaties such as Additional Protocol I, which define war crimes for which individuals may be held responsible and punished (Henckaerts 2000). In the past, however, punishment for war crimes depended largely on states' willingness to exercise "universal jurisdiction" under customary international law, or on establishment of ad hoc international criminal tribunals such as those for the Former Yugoslavia and Rwanda (ibid.). There has been discussion of a similar UN war-crimes tribunal for Sierra Leone, which could conceivably include plunder of natural resources within its jurisdiction (United Nations Security Council [UNSC] 2000).

Most recently, investigating the armed conflict in the DRC, a UN-appointed panel of experts recommended that the UN "Security Council consider establishing an international mechanism that will investigate and prosecute individuals" for criminal activities carried out during the conflict (Report of the Panel of Experts 2001). These violations include "economic criminal activities" such as illegal logging, mining, and shipping of natural resources from occupied areas of the DRC. These recommendations could be carried out in the larger context of a possible international tribunal to investigate all aspects of the DRC conflict, and indeed the United Nations Security Council Mission to the Great Lakes region has indicated that the UN is poised to assist the DRC in establishing such a tribunal (DRC Tribunal Possible 2001). The Panel of Experts report also recommended "establishing a permanent mechanism that would investigate illicit trafficking of natural resources in armed conflicts . . . such as those of Angola, the Democratic Republic of the Congo and Sierra Leone" (Para. 240).

The gaps in enforcement arising from the ad hoc enforcement of international law and the perceived need for a "permanent mechanism" have already led to creation of an International Criminal Court, which was chartered in mid-1998 but has not yet come into being. Indeed, the United States has been a highly vocal opponent of the Court, expressing concerns about subjecting U.S. peacekeeping and combat forces to the jurisdiction of an international tribunal. As discussed above, the Rome Statute of the International Criminal Court adopted the Protocol I prohibition against "widespread, long-term, and severe" environmental damage, but promptly weakened it by stipulating that such damage also must be "clearly excessive" in relation to the military advantage gained (Rome Statute, Article 8(2)(b)(iv)). Moreover, the prohibition applies only to international conflicts, not to civil wars or other internal strife, and appears to require that the defendant have intended to commit such damage (Austin and Bruch 1999). Nevertheless, the very act of positively declaring environmental damage to be a "war crime" and establishing a permanent forum for its prosecution constitutes a major step forward. While granting that the Court's initial cases will be closely watched, and necessarily focused on serious human rights violations, some commentators are already looking forward to future opportunities for the Court to expand its jurisdiction over environmental war crimes (Drumbl 2000).

A separate strategy is to focus not on the International Criminal Court's jurisdiction over enumerated war crimes, but on its broader mandate to prosecute "genocide" and "crimes against humanity" (Rome Statute, Article 5). Peter Sharp has argued that the worst acts of environmental degradation are inherently inseverable from these core crimes, and that where ecosystem damage results in the systematic poisoning or displacement of a civilian population, it may well rise to the level of genocide or crimes against humanity (Sharp 1999). One interesting result of this theory is that it does not necessarily require the existence of an international armed conflict, and arguably could apply even to peacetime situations where the stated goal of the harmful activity is resource extraction or other forms of economic development. Thus, where conflicts arise between exploitation of natural resources and the rights and livelihoods of native peoples–as in recent highly publicized cases in Nigeria, Myanmar, Ecuador, and Colombia–the International Criminal Court ultimately could provide an additional forum for investigating, prosecuting, and resolving these cases.

Civil Liability and Compensation

In addition to proposals that individuals be held criminally responsible for wartime actions that damage the environment, recent developments suggest a more general willingness to hold nations responsible and civilly liable for violations of the law of war, on theories similar to well-established tort and claims law. The United Nations Compensation Commission (UNCC), created after the 1990-91 Persian Gulf War by UN Security Council Resolutions 687 and 692, provides a model for how one nation can be required to compensate another for its environmental, human, and business losses (UNSC 1991a, 1991b). In brief, Resolution 687 presumed Iraq's liability "under international law for any direct loss, damage, *including environmental damage and the depletion of natural resources*, or injury to foreign governments, nationals, and corporations, as a result of Iraq's unlawful invasion and occupation of Kuwait" (emphasis added). Iraq also was declared liable for any damage caused by coalition forces in driving out its occupying army.

With liability presumed, the UNCC focuses exclusively on assessing the amount of personal, business, infrastructure, and environmental damages arising from the Iraqi invasion and occupation. Monies awarded by the commission come from the United Nations Compensation Fund, which receives 30% of the proceeds from UN-sanctioned Iraqi oil sales, as well as a portion of Iraq's frozen international assets. Still, it is unlikely that the available revenues will be sufficient to pay all of the claims. Nearly 100 nations submitted more than 2.6 million claims with an asserted value in excess of US$300 billion, with over US$40 billion of these claims being for environmental damage to groundwater, marine and coastal resources, terrestrial resources, and public health. Although the vast majority of claims have been processed and over US$15 billion awarded, the UNCC is only now starting to adjudicate the environmental claims. The first such decision awarded US$243 million to five Gulf countries to study and monitor the environmental and health damages they have incurred–a substantial amount, but less than a quarter of what had been sought ("UN to Pay $243 Mln for Gulf War Environment Studies" 2001).

While it is indisputable that Iraq's military actions caused severe ecological and public health damages, the precise extent of those damages remains unclear. There are few baseline ecological data available, and pre-war oil development activities caused earlier environmental impacts that were similar in kind, albeit much less extensive. Ongoing oil production activities will continue to affect the Gulf environment, masking some of the long-term environmental impacts of Iraq's actions. Additionally, it has proven difficult to assess accurately long-term ecological and public health impacts, again due in part to the challenges in isolating those impacts that resulted directly from wartime actions.

In the DRC, the report of the UN Panel of Experts also raised the possibility of developing an international commission to award monetary compensation for illegal logging and mining (Report of the Panel of Experts 2001). The panel recommended that individuals should receive compensation for damage or looting to their livestock, crops, and other property, and that the "governments of Burundi, Rwanda and Uganda and their allies should pay compensation to the companies whose properties and stocks of coltan, cassiterite, gold, timber and other materials were confiscated or taken . . ." (Para. 236). The panel also suggested that the responsible governments should be held liable for damage to wildlife in four national parks (Para. 237).

The prospect of extending the UNCC's Gulf War precedent to the conflict in the DRC presents a number of challenges. First, the likely defendants–the governments of Burundi, Rwanda, and Uganda, as well as the various rebel factions–are not likely to have substantial assets with which to pay a judgment against them. Unlike the case of Iraq, which remains oil-rich even as it struggles under international sanctions, there is no steady stream of governmental revenue that may provide a pool of money from which to satisfy claims. Rather, these countries are already saddled with significant debt burdens, and a judgment against them for the full amount of damages could further drain their economies. Moreover, much of the ill-gotten wealth from the DRC flowed to private individuals rather than the governments, posing a difficult question of tracing the profits and exercising jurisdiction over persons as well as governments. Nevertheless, a failure to compensate at least partially the people of the DRC for the looting of natural resources and wanton killing of wildlife could establish a dangerous incentive for future conflicts.

The destruction of natural resources in Colombia and Venezuela provides another model for civil compensation. As discussed above, guerrilla attacks on Colombian pipelines have spilled more than two million barrels of crude oil and contaminated Colombian waterways with aquatic impacts that extended to Venezuela (Rodriguez 1998). Rather than seeking compensation for damage to all environmental values, Venezuela reached an agreement with Colombia whereby ECOPETROL (Empresa Colombiana de Petróleos), Colombia's national petroleum company, compensates the Venezuelan petroleum company PDV (Petróleos de Venezuela) for the actual costs of cleaning up downstream contamination in Venezuela. While this agreement covers remediation costs, it does not include compensation for more extensive environmental or natural resources damages. This less formal model, based on negotiation, may represent a more practical approach, particularly in the typical situation of an ongoing internal armed conflict where there is no "deep-pocket" state defendant to hold liable.

Similarly, the twenty-fifth anniversary of the United States' withdrawal from Vietnam has generated a growing debate about the possibility of the U.S. government compensating the Vietnamese people for birth defects and other harms allegedly caused by spraying of Agent Orange (Dreyfuss 2000). While such compensation will be politically difficult to obtain in the face of opposition from the Pentagon and Congress, it could be modeled on the hard-won Veterans Administration policy of compensating U.S. soldiers who fought in Vietnam and now suffer from specific ailments or birth defects of their children that are attributable to Agent Orange. Alternatively, it has been suggested that the Vietnamese people could seek compensation from the defoliant makers, similar to the US$180 million judgment won by U.S. veterans against Dow Chemical Company and Monsanto Corporation.

Guidelines for Military, Peacekeeping, and Humanitarian Activity

The growing body of international law-of-war and environmental norms is reflected in legal guidelines and operational manuals produced by militaries themselves. These guidelines both incorporate treaty norms and interpret and extend them where the treaty language is broad or imprecise. As seen above, many law-of-war treaties expressly require signatory nations to undertake such dissemination and implementation efforts, and to educate and train their militaries accordingly. While some critics have expressed skepticism about the ultimate value of such self-regulation (Falk 2000), presently it is the main check on military discretion, absent a more comprehensive international regime of treaty enforcement.

In the United States, each branch of the military has its own operational law handbook that governs military actions during wartime, incorporating environmental considerations when determining lawful targets and weapons choice (Westing 2000; Quinn et al. 2000). Canada, Australia, Great Britain, and Germany similarly have military handbooks that require commanders to take environmental matters into consideration during wartime (Westing, 2000). International and multilateral forces also are starting to incorporate environmental norms into the military manuals that they follow (Westing 2000). To facilitate the further development of military guidelines, the International Committee of the Red Cross has published model guidelines for military manuals to incorporate environmental considerations during armed conflict (ICRC 1995).

Given the wide range of military operations other than full-scale war, domestic military manuals also have sought to address environmental concerns associated with peacetime training operations, police actions, and other activities that do not rise to the level of formally declared wars (Quinn et al. 2000). Likewise, the UN secretary general, through a 1999 bulletin entitled "Observance by United Nations Forces of International Humanitarian Law," explicitly adopted for UN peacekeeping efforts the same language found in the customary law of war, Protocol I, ENMOD, and the Rome Statute: "The United Nations force is prohibited from employing methods of warfare which may cause superfluous injury or unnecessary suffering, or which are intended, or may be expected to cause, widespread, long-term and severe damage to the natural environment." While it will be necessary to clarify the extent and application of this provision, it is an important milestone signaling that international institutions must incorporate environmental concerns into their operations.

Along the same lines, international operations that aim to meet the basic needs of refugees fleeing armed conflict also would benefit from operational guidelines regarding the environment. For example, to ameliorate and minimize environmental impacts of refugees, the UNHCR has developed and circulated policy and guidelines that explicitly address environmental concerns in the siting, management, and remediation of refugee and displaced persons camps (Jacobsen 1994). These guidelines would seek, among other things, to control deforestation around refugee camps, maintain clean camps, and alleviate pressure on water resources.

Import Bans and Trade Embargoes

Increasingly, individual nations and the international community are exploring broader legal options, such as import bans and trade embargoes, for cases where wars are being financed by illicit extraction of natural resources from tropical forests or where such extraction prolongs armed conflict. These mechanisms, for example the bans on Burmese timber or Iraqi oil, can address financing both of military activities, and of regimes generally deemed to be oppressive.

For example, the UN Security Council has sought to stem the flow of Angolan gray-market diamonds that has provided UNITA with the means to pay its troops and purchase weapons, fuel, and other material. In order to hasten peace in the country, the Security Council first imposed a weapons and fuel embargo in 1993 (UNSC 1993), which proved ineffective due to the steady flow of diamonds from UNITA. Consequently, in 1998, the Security Council passed Resolution 1173, which provided in paragraph 12(b) that "all States shall take the necessary measures . . . to prohibit the direct or indirect import from Angola to their territory of all diamonds that are not controlled through the Certificate of Origin regime of the [Angolan Government]." Furthermore, paragraph 12(c) prohibits the "sale or supply of equipment . . . used in mining or mining services" to people and entities in areas not controlled by the Angolan government. In imposing this import ban and embargo, the Security Council acted pursuant to Article 41 of the United Nations Charter, which empowers the Security Council to call upon nations to apply "measures not involving the use of armed force," with the underlying purpose of maintaining and restoring international peace.

The Angolan diamond ban has faced many problems, as certain outside individuals and nations have profiteered by skirting the ban. These include some African nations responsible for converting the diamonds into weapons and fuel, Bulgarian arms dealers, and Belgian diamond traders who recognize the distinctive, high-quality Angolan diamonds, but fail to ensure that the diamonds have the required certificate of origin. In 1999, the UN Security Council established a panel of experts to investigate and report on violations of sanctions imposed against UNITA with respect to the sale of arms and material, as well as how UNITA was able to circumvent the bans on UNITA petroleum and diamonds (UNSC 1999). In an attempt to shame the parties breaking the ban, the UN commission singled out particular individuals such as President Gnassingbe Eyadema of Togo and President Blaise Compaore of Burkina Faso (Fowler 2000). While it remains to be seen whether this tactic will work, *The Economist*, 18 March 2000, noted that similar tactics "had already frightened the DeBeers cartel into refusing to buy diamonds from Angola or other war zones." However, in a later article, 3 June 2000, it observed that these diamonds are simply being redirected to non-European markets, and concluded that "more concerted international action is needed."

In other recent cases, the international community has considered similar mechanisms for hastening peace and stability. As diamonds became a primary element in the financing and prolonging of wars not only in Angola, but also in Sierra Leone and the DRC–and as international efforts to stem the flow of these "conflict diamonds" sputtered–the United Nations General Assembly in January 2001 adopted Resolution 55/56. The purpose of which is to encourage nations to develop and implement a "simple and workable international certification scheme for rough diamonds" (UNGA 2001). Less than two months later, eighty U.S. Representatives (including both Democrats and Republicans) introduced H.R. 918 "to prohibit the importation of diamonds unless the countries exporting the diamonds into the United States have in place a system of controls on rough diamonds." While this bill only addresses diamonds, it could provide a precedent for establishing certification schemes for tropical timber and other natural resources that fuel armed conflicts around the world.

Indeed, the United States has sought a similar ban on timber from Liberia ("US Urges UN to Ban Liberian Diamonds and Timber" 2001). Liberia's largest timber company reportedly has played a key role in shipping arms to the Revolutionary United Front in Sierra Leone, in violation of the UN embargo ('The Role of Liberia's Logging Industry" 2001). However, while the diamond embargo continues to gain momentum, Ivan Watson reported on National Public Radio on 11 April 2001 that the UN Security Council at the last minute dropped a proposed timber embargo from its resolution at the behest of France and China, who together purchased 45% of Liberia's timber exports. Meanwhile, there continue to be calls for embargoes to prevent the looting of natural resources during armed conflict in other places, such as the DRC (Report of the Panel of Experts, 2001). In April 2001, the UN Panel of Experts recommended that the "UN Security Council should immediately declare a temporary embargo on the import or export of coltan, niobium, pyrochlore, cassiterite, timber, gold and diamonds from or to Burundi, Rwanda and Uganda" (Para. 221). To facilitate this embargo, the panel recommended, among other steps, the development of certification schemes for diamonds, timber, and non-timber forest products (Paras. 231-235). The diamond certification scheme seems to be moving forward, although a certification scheme for forest products has yet to gain much international support.

In short, the full potential of embargoes has yet to be explored. Yet even though official action often falls short, various international and non-governmental organizations have been successful in enacting, implementing, and monitoring a logging ban against Cambodia, where civil unrest and corruption has led to uncontrolled and illegal logging of tropical hardwoods (Talbott 1998). This followed earlier, unsuccessful efforts by the UN Security Council to ban all timber from Cambodia that were unsuccessful due to smuggling through Thailand to Japan (Peters 1999; Wolf 1996). Similar activism will continue to be necessary to fully implement the diamond and/or timber bans and certification schemes in Africa.

Emergency Remediation

Finally, it has been suggested that another solution to wartime environmental damage is to focus less on the actors than on the affected ecosystems themselves. Unlike criminal and civil liability, import bans, and trade embargoes–all of which depend on some finding of individual, group, or state culpability–this approach would approximate a "no-fault" or insurance-like system of emergency environmental remediation that could supplement the other approaches. Without giving up on the goal of seeking to apportion responsibility and to collect compensation wherever possible, a no-fault system acknowledges the reality that in many wartime instances, it will be politically and practically difficult to investigate, adjudicate, and enforce claims against individual and state defendants. Further, as in other kinds of environmental disasters, it is often more efficient from both an ecological and an economic point of view to mitigate damages as early as possible, and to reserve questions about responsibility, liability, and compensation for later.

Most proposals center on establishing some form of remediation fund–an "international Superfund" that could be built up from ongoing taxes on standing militaries or weapons sales. This fund could be drawn upon to clean up environmental damage in the immediate aftermath of armed conflict, and replenished by seeking indemnification from responsible parties where possible (Miller 2000; Austin and Bruch 1999). There are several precedents for such a fund in peacetime international environmental law, most notably the United Nations Convention on the Law of the Sea and its related liability conventions, which address spills of oil and other pollutants on the high seas (Mensah 2000). A similar liability and compensation scheme recently was added to the Basel Convention, though negotiators declined to include an emergency fund to address cases where the responsible party is unknown or insolvent. Both the World Heritage Convention and the Second Protocol to the Cultural Property Convention contain language that calls for the establishment of funds to advance their respective purposes. In the former case, this includes a "reserve fund" against disasters or natural calamities (World Heritage Convention, Articles 15-18, 21). In the latter case, the express goal "to provide financial or other assistance in relation to emergency, provisional or other measures to be taken in order to protect cultural property during periods of armed conflict or of immediate recovery after the end of hostilities" (Second Protocol, Article 29).

As with all such financing efforts, the key question will be how to generate and sustain a sufficient pool of funds to address incidents as they occur. It has been observed that the political will to create and endow new international institutions seems to be at an all-time ebb (e.g., Caron 2000). For example, while the UN in 1997 established a Trust Fund for Preventive Action Against Conflicts to much acclaim, it has received little actual funding (Renner 1999). The voluntary nature of the funds established by the World Heritage Convention and the Second Protocol to the Cultural Property Convention suggest that they may suffer a similar fate. Still, while not an all-encompassing solution, the idea of a fund is especially useful for the kinds of internecine conflict that is prominent in tropical regions, where it is often difficult to hold parties liable. Moreover, it is well suited for cases of environmental damage caused by refugee populations, which are largely composed of innocent third-party actors who should not be held accountable in a fault-based regime.

CONCLUSION

War's impacts on forests and ecosystems rich in natural resources are different, and in some ways more extensive, than its impacts on cities, industrial facilities, and other "traditional" military targets. Yet the existing legal mechanisms for addressing both situations are essentially the same, focusing on customary notions of military necessity, proportionality, and discrimination, and to a lesser extent on prevention of "widespread, long-term, and severe" damage. Such approaches may be appropriate for the first category of impacts discussed here, namely deliberate or indiscriminate targeting of the environment. They are less appropriate for the other three categories, where the damage is not caused directly by military forces in combat, but by the extractive uses made by government or rebel groups, refugee populations, or civilian populations at large. Modern warfare demands both a more systematic application of existing law-of-war norms to environmental damage, especially in internal conflicts, and an expanded array of legal tools and forums for addressing the more indirect forms of environmental damage. The very prevalence of post-conflict environmental assessments and calls for remediation, in situations ranging from the Gulf War to Kosovo to the recent African conflicts, suggests that the time is right for moving from ad hoc discussions to long-term solutions.

The various proposals examined in this article represent realistic steps in this direction. They have the further advantage of being attainable through incremental extensions of existing or emerging legal theories and institutions. Above all, adoption of the IUCN Draft Convention or a similar instrument would expressly import ecological conservation concerns into the law of war, especially if linked to a strong enforcement regime like that found in the Second Protocol to the Cultural Property Convention and then incorporated into military guidelines. Similarly, the continuing evolution of criminal and civil liability for wartime environmental damage should help to deter not only military forces, but also government and civilian actors who use conflict as cover for environmental despoliation. Finally, supplemental approaches, such as embargoes and bans or an emergency remediation fund, can be employed in cases where it is not feasible directly to halt, punish, or receive compensation from responsible parties. If developed in concert and used in appropriate cases, each of these mechanisms could contribute to a comprehensive regime for preventing, punishing, and redressing wartime damage to tropical forests.

REFERENCES

Austin, J. and C. Bruch. 1999. The Greening of Warfare: Developing International Law and Institutions to Limit Environmental Damage During Armed Conflict. Environmental Law Institute, Washington, DC.

Austin, J.E. and C.E. Bruch (eds.). 2000. The Environmental Consequences of War: Legal, Economic, and Scientific Perspectives. Cambridge University Press, Cambridge.

Boutwell, J. and M.T. Klare. 2000. A Scourge of Small Arms. Scientific American 280(6):51.

Bruch, C. 1998a. Addressing Environmental Consequences of War: Background Paper for the First International Conference on Addressing Environmental Consequences of War: Legal, Economic, and Scientific Perspectives. Environmental Law Institute, Washington, DC.

Bruch, C. 1998b. Law (And Lawyers) On The Battlefield. Environmental Forum 15(6):34.

Bruch, C.E. and J.E. Austin. 2000. The 1999 Kosovo Conflict: Unresolved Issues in Addressing the Environmental Consequences of War. Environmental Law Reporter 30:10069-79.

Burhenne, W.E. 1997. The Prohibition of Hostile Military Activities in Protected Areas. Environmental Policy and Law 27:373.

Caron, D.D. 2000. The Place of the Environment in International Tribunals. pp. 250-263 in J.E. Austin and C.E. Bruch (eds.). The Environmental Consequences of War: Legal, Economic, and Scientific Perspectives. Cambridge University Press, Cambridge.

Colombia Urges U.N. to Designate Bombing of Pipelines as Environment Treaty Violation. 1998. International Environmental Reporter (BNA) 21:175.

Colombian Army Cites Ecosystem Damage from Armed Left-Wing, Right-Wing Groups. 2001. International Environment Reporter (BNA) 24:57.

[Colombian] Ministerio del Medio Ambiente, 1998. "Se suspende descontaminación por la guerrilla hasta tanto cecen los ataques": Verano de la Rosa.

Convention on Prohibitions or Restrictions on the Use of Certain Conventional Weapons Which May Be Deemed to be Excessively Injurious or to have Indiscriminate Effects and Protocols. 1980. International Legal Materials 19:1524.

Convention on the Prohibition of Military or Any Other Hostile Use of Environmental Modification Techniques. 1976. U.S.T. 31:333; T.I.A.S. No. 9614; International Legal Materials 16:88.

DRC Tribunal Possible. TOMRIC, May 25, 2001. Available from World Wide Web: http://www.globalpolicy.org/security/issues/congo/2001/05court.htm.

Daltry J., and F. Momberg. 2000. Conservation News: The Cardamom Mountains Report. Oryx 34(3):227-228.

Dreyfuss, R. 2000. Apocalypse Still. Mother Jones (Jan.-Feb. 2000):42-51, 90.

Drumbl, M.A. 2000. Waging War Against the World: The Need to Move from War Crimes to Environmental Crimes. pp. 620-646 in J.E. Austin and C.E. Bruch (eds.). The Environmental Consequences of War: Legal, Economic, and Scientific Perspectives. Cambridge University Press, Cambridge.

Falk, R. 2000. The Inadequacy of the Existing Legal Approach to Environmental Protection in Wartime. pp. 137-155 in J.E. Austin and C.E. Bruch (eds.). The Environmental Consequences of War: Legal, Economic, and Scientific Perspectives. Cambridge University Press, Cambridge.

Fowler, R.R. 2000. Final Report of the UN Panel of Experts on Violations of Security Council Sanctions Against Unita. UN Doc. S/2000/203. Available from World Wide Web: http://www.globalpolicy.org/security/sanction/angola/report.htm.

Girot, P.O. and B.Q. Nietschmann. 1992. The Río San Juan. National Geographic Research and Exploration 8:52.

Hall, B. and D. Faber. 1989. El Salvador: Ecology of Conflict. Environmental Project on Central America, San Francisco.

Hay, A.W.M. 2000. Defoliants: The Long-Term Health Implications. pp. 402-425 in J.E. Austin and C.E. Bruch (eds.). The Environmental Consequences of War: Legal, Economic, and Scientific Perspectives. Cambridge University Press, Cambridge.

Henckaerts, J-M. 2000. International Legal Mechanisms for Determining Liability Under International Humanitarian Law. pp. 602-619 in J.E. Austin and C.E. Bruch (eds.). The Environmental Consequences of War: Legal, Economic, and Scientific Perspectives. Cambridge University Press, Cambridge.

ICRC (International Committee of the Red Cross). 1995. Law of War: Prepared for Action: A Guide for Professional Soldiers. ICRC, Geneva.

IUCN (International Union for Conservation of Nature and Natural Resources). 1998. Draft Convention on the Prohibition of Hostile Military Activities in Protected Areas. Gland, Switzerland.

Jacobsen, K. 1994. The Impact of Refugees on the Environment: A Review of the Evidence. Refugee Policy Group, Washington, DC.

Kim, K.C. 1997. Preserving Biodiversity in Korea's Demilitarized Zone. Science 278:242.

McNeely, J.A. 2000. War and Biodiversity: An Assessment of Impacts. pp. 353-378. J.E. Austin and C.E. Bruch (eds.). The Environmental Consequences of War: Legal, Economic, and Scientific Perspectives. Cambridge University Press, Cambridge.

Mensah, T.A., 2000. Environmental Damages Under the Law of the Sea Convention. pp. 226-249 in J.E. Austin and C.E. Bruch (eds.). The Environmental Consequences of War: Legal, Economic, and Scientific Perspectives. Cambridge University Press, Cambridge.

Miller, J.G., 2000. Civil Liability for War-Caused Environmental Damage: Models from United States Law. pp. 264-293 in J.E. Austin and C.E. Bruch (eds.). The Environmental Consequences of War: Legal, Economic, and Scientific Perspectives. Cambridge University Press, Cambridge.

National Academy of Sciences. Committee on the Effects of Herbicides in Vietnam. National Research Council. 1974. The Effects of Herbicides in Vietnam, Part B: Working Papers. Washington, DC.

Navarro, R.A. 1998. The Environmental Consequences of War: The Case of El Salvador. Unpublished document delivered at the First International Conference on Addressing Environmental Consequences of War (on file with authors). Washington, DC. June 10-12, 1998.

Peters, J.L. 1999. The Illegal Trafficking of Timber in Cambodia. Colorado Journal of International Environmental Law and Policy YB:102-111.
Protocol Additional (I) to the Geneva Conventions of August 12, 1949, and Relating to the Protection of Victims of International Armed Conflicts. 1977. International Legal Materials 16:1391.
Protocol Additional (II) to the Geneva Conventions of August 12, 1949, and Relating to the Protection of Victims of Non-International Armed Conflicts. 1977. International Legal Materials 16:1442.
Quinn, J.P. et al. United States Navy Development of Operational-Environmental Doctrine. pp. 156-170 in J.E. Austin and C.E. Bruch (eds.). The Environmental Consequences of War: Legal, Economic, and Scientific Perspectives. Cambridge University Press, Cambridge.
Renner, M. 1999. Ending Violent Conflict. Worldwatch Institute, Washington, DC.
Report of the Panel of Experts on the Illegal Exploitation of Natural Resources and Other Forms of Wealth of the Democratic Republic of the Congo. 2001. UN Doc. S/2001/357. Available from World Wide Web: http://www.un.org/Docs/sc/letters/2001/357e.pdf.
Rodriguez, A.J. 1998. The Undeclared War for Control over Natural Resources: Armed Conflict and Environmental Degradation in Colombia. Unpublished document delivered at the First International Conference on Addressing Environmental Consequences of War (on file with authors). Washington, DC. June 10-12, 1998.
The Role of Liberia's Logging Industry on National and Regional Insecurity. IRIN, January 24, 2001. Available from World Wide Web: http://www.globalpolicy.org/security/issues/diamond/2001/0124gwit.htm.
Rome Statute of the International Criminal Court. U.N. Doc. A/CONF.183/9 (1998).
Savitz, D.A. et al. 1993. Vietnamese Infant and Childhood Mortality in Relation to the Vietnam War. American Journal of Public Health 83:1134.
Sayagues, M. 1999. Angola: Losing Trees to War. InterPress Service, September 22, 1999.
Schecter, A. et al. 1995. Agent Orange and the Vietnamese: The Persistence of Elevated Dioxin Levels in Human Tissues. American Journal of Public Health 85:516.
Schmidt, J. 1999. Soldiers in the Gorilla War. International Wildlife 29(1):12-21.
Schmitt, M.N., 2000. War and the Environment: Fault Lines in the Prescriptive Landscape. pp. 87-136 in J.E. Austin and C.E. Bruch (eds.). The Environmental Consequences of War: Legal, Economic, and Scientific Perspectives. Cambridge University Press, Cambridge.
Sharp, P. 1999. Prospects for Environmental Liability in the International Criminal Court. Virginia Environmental Law Journal 18:217-243.
Talbott, K. 1998. Logging in Cambodia: Politics and Plunder. In: F.Z. Brown and David G. Timberman (eds.). Cambodia and the International Community: The Quest for Peace, Development, and Democracy. Asia Society, New York. Available from World Wide Web: http://www.asiasociety.org/publications/cambodia/logging.html.
Tarasofsky, R.G. 2000. Protecting Specially Important Areas During International Armed Conflict: A Critique of the IUCN Draft Convention on the Prohibition of Hostile Military Activities in Protected Areas. pp. 567-578 in J.E. Austin and C.E.

Bruch (eds.). The Environmental Consequences of War: Legal, Economic, and Scientific Perspectives. Cambridge University Press, Cambridge.

UN to Pay $243 Mln for Gulf War Environment Studies. Reuters, June 21, 2001. Available from World Wide Web: http://www.enn.com/news/wirestories/2001/06/06212001/reu_gulf_44073.asp.

UNESCO (United Nations Educational, Scientific and Cultural Organization). 1999. Second Protocol to the Hague Convention of 1954 for the Protection of Cultural Property in the Event of Armed Conflict.

UNESCO. 1972. Convention Concerning the Protection of the World Cultural and Natural Heritage (World Heritage Convention). UN Treaty Series No. 15511.

UNESCO. 1954. Convention for the Protection of Cultural Property in the Event of Armed Conflict. UN Treaty Series No. 3511.

United Nations General Assembly. 2001. Resolution 55/56. UN Doc. A/RES/55/56.

United Nations Security Council (UNSC). 2000. Resolution 1315. UN Doc. S/RES/1315.

UNSC. 1999. Resolution 1237. UN Doc. S/RES/1237.

UNSC. 1998. Resolution 1173. UN Doc. S/RES/1173.

UNSC. 1993. Resolution 864. UN Doc. S/RES/0864.

UNSC. 1991a. Resolution 687. UN Doc. S/RES/0687.

UNSC. 1991b. Resolution 692. UN Doc. S/RES/0692.

U.S. Senate Committee on Foreign Relations. 1972. Prohibiting Military Weather Modification.

US Urges UN to Ban Liberian Diamonds and Timber. Reuters, 18 January 2001. Available from the World Wide Web: http://www.globalpolicy.org/security/issues/diamond/2001/0119lib.htm.

Vöneky, S. 2000. Peacetime Environmental Law as a Basis of State Responsibility for Environmental Damage Caused by War. pp. 190-225 in J.E. Austin and C.E. Bruch (eds.). The Environmental Consequences of War: Legal, Economic, and Scientific Perspectives. Cambridge University Press, Cambridge.

Waldman, P. 1997. Babies of Cam Nghia Village [online]. Dow Jones & Co., Inc. Available from World Wide Web: http://www.mayerson.com/agentorange/ao4.html.

Weinberg, B. 1991. War on the Land: Ecology and Politics in Central America. Zed Press, London.

Weinberg, B. 1999. Colombia's Endless Nightmare. Native Americas 16(2):49-57.

Weinberg, S. 1997. Civil War and the Environment in El Salvador [online]. Available from World Wide Web: http://gurukul.ucc.american.edu/TED/zelsalv.htm.

Weiskopf, J. 1999. Aerial Fumigation Continues to Draw Complaints of Environment, Health Effects. International Environmental Reporter (BNA) 22:690.

Westing, A.H. 2000. In Furtherance of Environmental Guidelines for Armed Forces during Peace and War. pp. in 171-181. J.E. Austin and C.E. Bruch (eds.). The Environmental Consequences of War: Legal, Economic, and Scientific Perspectives. Cambridge University Press, Cambridge.

Westing, A.H. 1998-99. Transfrontier Reserve for Peace and Nature on the Korean Peninsula. International Environmental Affairs 10(1): 8-17.

Westing, A.H. 1977. Weapons of Mass Destruction and the Environment. Taylor & Francis, London, Philadelphia.
Westing, A.H. 1976. Ecological Consequences of the Second Indochina War. Almqvist & Wiksell, Stockholm.
Whaley, F. 2000. Search for the Hidden Killers. International Wildlife 30(2):24-30.
Wolf, H.A. 1996. Deforestation in Cambodia and Malaysia: The Case for an International Legal Solution. Pacific Rim Law & Policy Journal 5:429.

List of Acronyms

ADB	Asian Development Bank
AFDL	Alliance des Forces Démocratiques pour la Libération du Congo-Zaire; Alliance of the Democratic Forces for the Liberation of Congo-Zaire
APHI	Asosiasi Pengusaha Hutan Indonesia; Association of Indonesian Forest Concessionaires
APKINDO	Asosiasi Panel Kayu Indonesia; Indonesian Wood Panel Association
AWF	African Wildlife Foundation
CAA	Congress of the Armies of the Americas
CCW	Convention on Certain Conventional Weapons
CGI	Consultative Group on Indonesia
CIAV-OAS	Comisión Internacional de Apoyo y Verificación–OAS; International Support and Verification Commission–Organization of American States
CIMI	Conselho Indigenista Missionário; Indigenist Missionary Council
DFGF	Dian Fossey Gorilla Fund International
DMZ	Demilitarized Zone
DRC	Democratic Republic of the Congo
ECCP	European Centre for Conflict Prevention
ECOPETROL	Empresa Colombiana de Petróleos
ELN	Ejército de Liberación Nacional; National Liberation Army (Colombia)
ENMOD	Convention on the Prohibition of Military or Any Other Hostile Use of Environmental Modification Techniques
EU	European Union

FAR	Forces Armées Rwandaises; Rwandan Armed Forces
FAR	Fuerzas Armadas Revolucionarias; Revolutionary Armed Forces (Nicaragua)
FARC	Fuerzas Armadas Revolucionarias de Colombia; Revolutionary Armed Forces of Colombia
FDN	Fuerza Democrática Nicaragüense; Nicaraguan Democratic Force
FEA	Frente Ecológico Armado; Armed Ecological Front (Nicaragua)
FFI	Fauna and Flora International
FSLN	Frente Sandinista de Liberación Nacional; Sandinista Front for National Liberation
FUAC	Frente Unido Andrés Castro; Andrés Castro United Front (Nicaragua)
GDP	Gross Domestic Product
GIS	Geographic Information System
GNP	Gross National Product
GTZ	Gesellschaft für Technische Zusammenarbeit; German Technical Agency for Cooperation
HPH	Hak Pengusahaan Hutan
ICCN	Institut Congolais pour la Conservation de la Nature; Congolese Institute for the Conservation of Nature
ICG	International Crisis Group
IGCP	International Gorilla Conservation Programme
IMF	International Monetary Fund
INRA	Instituto Nicaragüense de Reforma Agraria; Nicaraguan Institute of Agrarian Reform
ITCI	International Timber Corporation of Indonesia
IUCN	International Union for Conservation of Nature–The World Conservation Union
IZCN	Institut Zairois pour la Conservation de la Nature; Zairian Institute for the Conservation of Nature
KKN	korupsi, kolusi, dan nepotisme; corruption, collusion, and nepotism
KRC	Karisoke Research Center (Rwanda)
LIPI	Lembaga Ilmu Pengetahuan Indonesia; Indonesian Institute of Science

MADENSA	Maderas y Derivadas de Nicaragua S.A.
MARENA	Ministerio del Ambiente y los Recursos Naturales; Ministry of Environment and Natural Resources (Nicaragua)
MINIREISO	Ministère de la Réhabilitation et de l'Intégration Sociale; Ministry of Rehabilitation and Social Integration (Rwanda)
MINITERE	Ministère des Terres, des Etablissements Humains, et de l'Environnement; Ministry of Lands, Resettlement, and Environment Protection (Rwanda)
MISURA	Miskitu Sumu Ramas; Miskitos, Sumos, Ramas
MISURASATA	Miskitu Sumu Rama Sandinista Asla Takanka; Miskitos, Sumos, Ramas, and Sandinistas Working Together
MPR	Majelis Permusyawaratan Rakyat; People's Consultative Assembly of Indonesia
NGOs	Non-Governmental Organizations
ONUCA	Grupo de Observadores de las Naciones Unidas para Centroamérica; United Nations Observer Group in Central America
ORTPN	Office Rwandais de Tourisme et des Parcs Nationaux; Rwandan Office for Tourism and National Parks
PCFN	Projet Conservation de la Foret de Nyungwe; Nyungwe Forest Conservation Project (Rwanda)
PDV	Petróleos de Venezuela
PNA	Parc National de l'Akagera; Akagera National Park (Rwanda)
PNKB	Parc National de Kahuzi-Biega; Kahuzi Biega National Park (Democratic Republic of the Congo)
PNV	Parc National des Volcans; Volcano National Park (Rwanda)
PNVi	Parc National des Virunga; Virunga National Park (Democratic Republic of the Congo)

POPOF	Pole Pole Foundation (Democratic Republic of the Congo)
RAAN	Región Autónoma Atlántico Norte; North Atlantic Autonomous Region (Nicaragua)
RCD	Rassemblement Congolais pour la Démocratie; Congolese Rally for Democracy
RCD-ML	RCD-Mouvement de la Libération; RCD-Liberation Movement
RCD-MLC	RCD-Mouvement pour la Libération du Congo; RCD-Congolese Liberation Movement
RN	Resistencia Nicaragüense; Nicaraguan Resistance
RPA	Rwandan Patriotic Army
RPF	Rwandan Patriotic Front
SLORC	State Law and Order Restoration Council
TGHK	Tata Guna Hutan Kesepakatan; Consensus Forest Land Use Plan (Indonesia)
TNC	The Nature Conservancy
TNI	Tentara Nasional Indonesia; The Indonesian Military
TRUBA	Tri Usaha Bhakti
UN	United Nations
UNCC	United Nations Compensation Commission
UNESCO	United Nations Educational, Scientific and Cultural Organization
UNF	United Nations Foundation
UNHCR	United Nations High Commissioner for Refugees
UNITA	União Nacional para a Independência Total de Angola; National Union for the Total Independence of Angola
UN-ODCCP	United Nations Office for Drug Control and Crime Prevention
UNSC	United Nations Security Council
USAID	United States Agency for International Development
UWA	Uganda Wildlife Authority
WCS	Wildlife Conservation Society
WWF	World Wide Fund for Nature
YATAMA	Yapti Tasbaya Maraska nani Asla Takanka

Appendix

Presentations Made at the International Conference:

War and Tropical Forests:
New Perspectives on Conservation in Areas of Armed Conflict

Held at Yale School of Forestry and Environmental Studies,
New Haven, Connecticut
March 31-April 1, 2000

Biodiversity, War, and Tropical Forests
Jeffrey McNeely (International Union for the Conservation of Nature–The World Conservation Union [IUCN])

Cambodia: Corruption, War, and Forest Policy
Patrick Alley (Global Witness)

Forests and Conflict: The Colombian Case
Manuel Rodriguez (National Environmental Forum; Colombia)

Legal Mechanisms for Addressing Wartime Damage to Tropical Forests
Jay Austin (Environmental Law Institute)

Forests in the Time of Violence: Conservation Implications of the Colombian War
María D. Álvarez (Columbia University)

The Role of the Military and the Fate of the Forests: Indonesia in the Context of a Rapidly Evolving Southeast Asian Political Economy
Kirk Talbott (Conservation International)

Restarting Conservation in Post-War Liberia
Jamison Suter (Fauna & Flora International)

The Impact of War on the Forests in El Salvador: Opportunities for Rehabilitation
Ricardo Navarro (Centro Salvadoreño de Tecnología Apropiada–CESTA)

Contras and Comandantes: Armed Movements and Forest Conservation in Nicaragua's Bosawas Biosphere Reserve in the 1990s
David Kaimowitz (Center for International Forestry Research–CIFOR)

Impacts of Armed Conflict on Forest Conservation Programs in Guatemala: Lessons Learned in the Maya Biosphere Reserve
Keith Kline (US Agency for International Development–USAID Guatemala)

The Zapatistas and Lacandon Selva: From Conservation to Counterinsurgency in the Chiapas Rainforest
Bill Weinberg (Journalist and Author)

Building Partnerships in the Face of Political and Armed Crisis: The International Gorilla Conservation Programme
Annette Lanjouw (International Gorilla Conservation Programme)

Lessons Learned from On-the-Ground Conservation in Rwanda and the Democratic Republic of the Congo
Andrew Plumptre (Wildlife Conservation Society–WCS)

Slaughter of Gorillas and Crisis of Conservation in the Kahuzi-Biega National Park, Democratic Republic of the Congo
Juichi Yamagiwa (Kyoto University)

Eastern Democratic Republic of the Congo: Why Should We Care?
Omari Ilambu (Yale School of Forestry and Environmental Studies)

Index

9781560220992

T - #0375 - 101024 - C4 - 212/152/13 - PB - 9781560220992 - Gloss Lamination